建筑立场系列丛书 No.22

建筑谱系传承
Genealogical Reasoning

中文版

韩国C3出版公社 | 编

于风军 杨惠馨 刘小玲 王平 徐雨晨 辛敏裕 郑海荣 | 译

大连理工大学出版社

建筑谱系传承

- 004 父与子_Aldo Vanini
- 010 卡斯卡伊斯城堡酒店_Gonçalo Byrne+João Alexandre Góis+David Sinclair
- 022 海德堡城堡游客中心_Max Dudler
- 030 Moka住宅_A-Cero
- 038 "代表"住宅_FORM/Kouichi Kimura Architects
- 048 栗树双体住宅_Lussi+Halter Partner AG
- 056 塞恩斯伯里实验室_Stanton Williams Architects
- 070 Riberas de Loiola的耶稣教堂_Rafael Moneo

建筑立场系列丛书 No.22

城市设计
从限制到优势

- 082 从限制到优势_Silvio Carta
- 088 Capelinhos火山讲解中心_Nuno Ribeiro Lopes Arquitectos
- 102 Fontinha码头_Alexandre Burmester Arquitectos Associados
- 116 圣伊莎贝尔住宅_Bak Gordon Arquitectos
- 128 安嫩代尔住宅_CO-AP
- 140 Ceschi住宅_Traverso-Vighy Architetti
- 150 Rizza住宅_Studio Inches Architettura
- 158 Zayas住宅_García Torrente Arquitectos
- 168 潜望镜式住宅_C+ Arquitectos

178 建筑师索引

Genealogical Reasoning

004 *Fathers and Sons_Aldo Vanini*

010 Cascais Citadel Hotel_Gonçalo Byrne+João Alexandre Góis+David Sinclair

022 Heidelberg Castle Visitor Center_Max Dudler

030 Moka House_A-Cero

038 House of Representation_FORM/Kouichi Kimura Architects

048 Chestnut Tree Twin Houses_Lussi+Halter Partner AG

056 Sainsbury Laboratory_Stanton Williams Architects

070 Iesu Church in Riberas de Loiola_Rafael Moneo

Urban How
Constraints to Blessings

082 *Over the Constraints_Silvio Carta*

088 Capelinhos Volcano Interpretation Center_Nuno Ribeiro Lopes Arquitectos

102 Fontinha Wharf_Alexandre Burmester Arquitectos Associados

116 Santa Isabel Houses_Bak Gordon Arquitectos

128 Annandale House_CO-AP

140 Ceschi House_Traverso-Vighy Architetti

150 Rizza House_Studio Inches Architettura

158 Zayas House_García Torrente Arquitectos

168 Periscope House_C+ Arquitectos

178 Index

建筑谱系传承

在建筑学领域,弗里德里希·尼采的"永劫回归观"是尤为有益的应用。所有的建筑理念和设计都必须符合任何人也无法逾越的静力学定律和根深蒂固的人类学空间概念。鉴于这两个前提,一代又一代建筑师之间存在着怎样千丝万缕的联系呢?连接着一代又一代的建筑师的那根线是什么呢?

过去,建筑知识和基本原则都是通过命令和论述来传播的,如同对现实中的整体概念和本体论概念的表述。现代特色和20世纪的科学弱化了这一观念,认为僵硬而正式的建筑模型是不可能延续的。

建筑师除了或多或少拥有复杂的形式创作上的灵感之外,还有父辈与子辈之间、前辈建筑师与后辈建筑师之间传承的传统建筑。以下建筑从不同方面向我们展示了建筑是如何承袭过去而发展演化的。

Friedrich Nietzsche's idea of eternal recurrence finds particularly useful application in the discipline of architecture. The necessity of operating within the insurmountable laws of statics and the established anthropological conception of space prevents any substantially rash moves and any abandonment of the fundamental principles of design. Given these premises, what is the relationship between the various generations of architects? What is the common thread connecting the evolution from one generation of architects to the next? In the past, the transmission of the knowledge and principles of architecture transpired through orders and treatises, as an expression of a monolithic and ontological conception of reality. Modernity and twentieth-century science have undermined this conception by rendering any continuity of rigid formal models impossible.

Beyond any more or less sophisticated formal inspiration, is the logical construction of the heritage that must be transmitted between fathers and sons, between the architects who precede and those who follow. The following examples represent some of the ways in which architecture can evolve without breaking continuity with the past.

卡斯卡伊斯城堡酒店
/Gonçalo Byrne+João Alexandre Góis+David Sinclair
海德堡城堡游客中心/Max Dudler
Moka住宅/A-Cero
"代表"住宅/FORM/Kouichi Kimura Architects
栗树双体住宅/Lussi+Halter Partner AG
塞恩斯伯里实验室/Stanton Williams Architects
Riberas de Loiola的耶稣教堂/Rafael Moneo
父与子/Aldo Vanini

Cascais Citadel Hotel
/Gonçalo Byrne+João Alexandre Góis+David Sinclair
Heidelberg Castle Visitor Center/Max Dudler
Moka House/A-Cero
House of Representation/FORM/Kouichi Kimura Architects
Chestnut Tree Twin Houses/Lussi+Halter Partner AG
Sainsbury Laboratory/Stanton Williams Architects
Iesu Church in Riberas de Loiola/Rafael Moneo
Fathers ans Sons/Aldo Vanini

父与子

每位建筑师十有八九都有父辈和子辈,因此说,除了建筑伊始,合理的建筑过程一直在循环往复,从不间断。"每件事物都反复存在了无数次,这是因为所有力的总体条件总是回归。"[1]

尽管形式的多重性使建筑形式多种多样,但建筑本身自有一套精确定义的游戏规则。建筑不仅是一门艺术或一套自我表达的独立体系,它还是必须遵循静力学定律、通过精密编码的人类学设计来满足一些基本需要的体系。因此,建筑知识和技艺的传播严格说来只有依靠保护才能得以实现。

事实上,许多世纪以来,建筑知识从一代到另一代的过渡都是严格通过命令和论述实现的。后代所拥有的自由就是重新整合从前辈大师那里学到的正规建筑元素,但有一种情况该另当别论,那就是像混凝土这样的新建筑材料出现的重大革命性时刻或关于力的分布人们获得了新的静力学知识。然而,在这两种情形下,建筑发展新的可能性也没能胜过人们为了与整个建筑教条保持一致而依循权威的、约定俗成的正规体系的需要。因此,建筑学科知识的传播基本上还是通过高级形式的模仿而实现的。

随着启蒙运动的到来,对建筑形式的模仿才逐渐被人们对建筑理论更深刻的理解所取代。19世纪上半叶,法国建筑师安托万•克特米瑞•狄•昆西基于"基本原则"这一概念创立了新的建筑理论,这一基本原则就是所有建筑形式都包含的逻辑元素。《建筑字典》[2]对"类型"的定义超出了循规蹈矩的模仿这一传统理念,依赖合理的建筑来指导建筑演变。维奥莱•勒•迪克[3]也持有类似观点。乔治•葛拉西认为[4],从命令性的僵硬死板的规定向"类型"过渡引入了一种基于真正"概念工具"的新方法。这是基于通用和理论知识开展建筑过程的开端,并最终形成了20世纪的建筑理性主义。

工业革命和资产阶级的兴起推动了大范围新建筑功能的出现,这些新建筑功能难以用传统正式的体系来阐释。号称三大革命性建筑师的勒杜、布雷和勒克设计了一系列建筑项目,这些建筑项目的空间、体量和形状主要取决于功能而非体现在对整体划分和古典因素的尊重。尽管"修复"的影响和更加传统的新古典主义建筑的回归削弱了这些建筑师的影响,但他们的理性主义和功能主义的愿景将再度出现在现代建筑运动中。

现代建筑运动的众多发起人引入了不同的新价值观和内涵并被其追随者们承袭接受[5],1929年,布鲁诺•陶特将其思想综合概括为五点。由亨利•范•德•威尔德的工艺学校所开创的这一"父与子"之间的新型关系被浓缩在了Gropius Bauhaus(格罗皮乌斯创立的包豪斯设

Fathers and Sons

For every architect it is possible to find a father and a son, given that, with the exception of the very first beginnings of architecture, the logical architectural process has been an endlessly circular path. *"Everything has existed countless times, as the overall condition of all the forces always returns"*.[1]

In spite of the multiplicity of forms by which it expresses itself, architecture plays its game with a precisely defined deck of cards. Architecture is not merely an art, an individual system of self-expression. Architecture is a system that answers to primary necessities by means of well coded anthropological schemes inside the mandatory laws of statics. With the matter thus framed, the transmission of knowledge and craft in architecture would seem possible only in strict terms of conservation.

In fact, for many centuries, the transition from one generation to another occurred strictly through orders and treatises. The only freedom offered to following generations was that of recombining formal elements learned from the masters. The only exceptions came in moments of great revolutionary conquests of new materials such as concrete, or via new knowledge of statics, as in the distribution of forces in the archivaulted system. In both these cases, however, the new possibilities did not overcome the need for canonical and established formal systems in accordance with a dogmatic idea of the world. Once again, the transmission of the discipline transpired more or less through advanced forms of imitation.

It was only with the Enlightenment that imitation of forms came to be progressively replaced by deeper understandings of the theoretical reasons for architecture. In the first half of the XIXth century, Antoine Quatremère de Quincy founded a new theory of architecture based on the concept of the "elementary principle", a logical element which precedes the forms that derive from it. The concept of "type" given in the *Dictionnaire d'Architecture*[2] definitively transcends the traditional idea of didactic imitation, as practiced in the orders, relying on logical construction to guide the evolution of architecture. We find a similar conceptualization in Viollet le Duc[3]. As Giorgio Grassi has maintained[4], this transition from the rigid rules of the orders to the "type" introduced a new approach based on a true "conceptual instrument". This was the starting point of an architectural process based on general and theoretical knowledge that would lead to XXth century rationalism. The Industrial Revolution and the rise of the bourgeoisie occasioned the advent of a large range of new building functions that could scarcely be interpreted under traditional formal systems. The so-called architects of the Revolution – Ledoux, Boullée and Lequeu – produced a repertoire of projects in which spaces, volumes and shapes were determined mainly by function rather than by a respect for partitions and classical elements. Although the lessons of these architects were diluted by the effects of the "Restauration" and by the return of the more traditional approach of Neoclassical architecture, their rationalist and functionalist vision would reemerge in the Modern Movement.

The lesson of the founding fathers of the Modern Movement, as synthesized in five points by Bruno Taut in 1929, introduces new

斯图加特魏森霍夫某住宅,德国,勒·柯布西耶,1926年
House at Weissenhof in Stuttgart, Germany by Le Corbusier, 1926

计体系)之中,并得到了很好地呈现和制度化。Gropius Bauhaus是建筑师基于方法而非形式代代延续传承的革新典范。教育和教学法是传播这一至关重要建筑经验的关键。在建筑理性主义和现代建筑运动时期,对年轻建筑师的教学和训练从没有被给予如此重视,20世纪上半叶一些伟大的建筑大师——勒·柯布西耶、弗兰克·劳埃德·赖特、沃尔特·格罗皮乌斯和密斯·凡·德·罗——也从来没有建筑师像他们这样被好几代建筑师竞相膜拜。

然而,为了区别于教条化的传统建筑理念,并更好地表达现代建筑运动,功能主义建筑方法最终形成了自己的语言,我们也普遍引用这些语言来描绘20世纪整个建筑领域。20世纪,在通用的功能主义概念指导下,再加上人们对社会历程的重视,建筑知识和技术传播的过程不再是线性的,而是分化成多种学派、多种运动,形成多种趋势。

尽管建筑领域的一些理论看起来模棱两可、零零碎碎,有有机主义、理性主义、建构主义、国际风格,也有新近由Scuola di Tendenza发起的运动和新的民族学派,但是我们总能找到主线。然而,随着后现代主义和解构主义的出现,随着人们越来越追求建筑的壮观和建筑师越来越追逐成为建筑明星的伟绩,要想找到一件正常一点的建筑作品变得越来越难。

自相矛盾的是,由后现代主义、弱势思想和解构主义等设计学派所提出的纯粹线性的危机甚至没有削弱基于逻辑学和方法论的后古典时期建筑思想的传播。

如今,找到代代建筑师和设计师之间延续承袭的可能性有多大?人们是否愿意使建筑理论原则保持生机与活力?抑或仅仅基于国际杂志或网络对正式的建筑模型的传播,模仿的做法是否将再次盛行?

由于21世纪全球化和文化多元性,工作室、学校或各种建筑运动都不能完全决定理想的父子相承关系。建筑研究始于各种各样的灵感,历史的、纲领性的、类似的,或仅仅是形式主义的,并吸收了众多的理论和实例,但尽管如此,正如弗里德里希·尼采所说,没有什么是真正原创的,一切都是回归,在当代面纱下隐藏的一切建筑思想和理论也不例外。

"代表"住宅/京都,日本_FORM/木村浩一建筑师事务所

木村浩一建筑师事务所设计的灰白色"代表"住宅,优雅、纯洁而且真实,从中不难看出设计师深受阿道夫·卢斯和包豪斯建筑理念的影响。不止这栋建筑如此,从该事务所的所有作品中都可以看到其通过纯洁的建筑对空间构成深刻的理解和思考。利用建筑的纯净、通过光线的穿透来创造空间的这种方法确立了建筑内部和外部之间的

and different values and content to be transmitted to disciples [5]. As inaugurated by Henry van de Velde's Kunstgewerbeschule, the epitome of this new relationship between "fathers and sons" would be well represented and institutionalized by Gropius Bauhaus, a revolutionary example of the continuity between generations of architects based on method rather than forms. Education and didactics would be the focal point of that crucial architectural experience. The teaching and training of young architects was never accorded so much importance as during the experience of Rationalism and the Modern Movement; never as in the first half of the last century, when the great masters – Le Corbusier, Frank Lloyd Wright, Walter Gropius and Mies van de Rohe – represented reference points and examples for entire generations of architects.

Nevertheless, this functionalist approach would eventually produce its own formal vocabulary, intended to be anticanonical and highly articulate, but to which we would universally refer for the majority of the XXth century. During that century, under the common concept of functionality and a great attention to social processes, the process of the transmission of architecture would no longer be linear, but would diverge into a multiplicity of schools, movements and trends.

Despite the ambiguous fragmentation of architecture into Organicism, Rationalism, Constructivism, the International Style or the more recent movements of the Scuola di Tendenza and the new national schools, one could always find the ideal leading wires. However, with the advent of Postmodernism and Deconstructivism, it would become ever more difficult to find the common matrices of a production increasingly interested in spectacularity and in the personal exploits of the archistars.

Paradoxically, the crisis of a solid linearity introduced by such schools of philosophical thought as Postmodernism, Weak Thought, and Deconstructivism ended up weakening even the post-classical transmission of architectural thought based on logics and method. What remains today of the possibility of finding a continuity between generations of architects and designers? Is there a desire to keep alive the theoretical principles of architecture, or does a process of imitation once again prevail, based merely on the dissemination of formal models via international magazines or through the Web?

Due to the globalization and multicultural complexity of the XXIst century, relationships of ideal fatherhood cannot be uniquely determined according to the traditional concepts of the workshop, school or movement. Research starts from a variety of inspirations, which may be historical, programmatic, analogical, or simply formalistic, drawing on an enormous theoretical and practical repertoire, but in which, as Friedrich Nietzsche warned, nothing is truly original and everything returns, even behind the mask of the contemporary.

House of Representation / Kyoto, Japan_FORM/Kouichi Kimura Architects

In the elegant purity and genuineness of the House of Representation's off-white volumes by Kouichi Kimura Architects it is pos-

圣母大教堂，美国洛杉矶，拉斐尔•莫尼奥，2002年
Cathedral of Our Lady of the Angels, Los Angeles, U.S.A. by Rafael Moneo, 2002

连续性，同时也给人情感上的亲密感，形式既不太抽象又不太单一，这一灵感来自于密斯•凡•德•罗的经历。

光与影的使用、建筑表面与建筑材料的契合、置身于周边的乡村景色之中，这一切使任何装饰都显得多余，正符合了露西安的至理名言"Ornament ist Verbrechen"——装饰是犯罪。每一个空间、空间中的每一个空隙都精确地表达了其在整体结构中的实用功能。整体结构的设计更多是基于空间关系而不是单一元素。这样的建筑在20世纪20年代斯图加特的魏森霍夫深受欢迎。

栗树双体住宅/卢塞恩，瑞士_Lussi+Halter Partner AG

栗树双体住宅雄伟的几何风格外观使人产生强烈的似曾相识之感，建筑师本身的国别更突出了这一感觉。

这不得不让人追忆起勒•柯布西耶设计的位于法国普瓦西的萨沃伊别墅那简易而不朽的建筑形式，回忆起它赋予建筑领域以规则与秩序那谜一般的存在。然而，萨沃伊别墅只是一个起点，是一个能够完成双体住宅楼项目正式而充满智慧的模型。萨沃伊别墅的内部空间摆脱了其规则的正方形外观限制，设计生成了许多小空间和内部景色。正如勒•柯布西耶的别墅，整个建筑都可以使用坡道贯穿。为了享受与周围乡村景色的私密亲近，勒•柯布西耶利用他的"激发诗意情感的客体"理论设计了露台，出于同样的考虑，游泳池也位于露台上。顺便说一下，用墙隔开两座房子，两座房子犹如两只眼睛在闪烁，这足以形成整体建筑物的对称，但勒•柯布西耶的设计中很少使用对称。建筑外部和内饰材料的使用也说明设计师对所参考的建筑模型的背离。

栗树双体住宅没有对建筑大师敬而生畏，而是机智地展示了再现建筑大师不朽主题的可能性，使之与环境契合，与当代契合。

塞恩斯伯里实验室/剑桥大学，英国_Stanton Williams建筑师事务所

塞恩斯伯里实验室应归功于科学领域和建筑领域许多先辈们所付出的努力。

实验室坐落于剑桥大学植物园之内，植物园由剑桥大学钦定讲座教授、植物学教授、查尔斯•达尔文的良师益友约翰•斯蒂文斯•亨斯洛牧师设计建造。

从实验室的建筑外观我们可以看到对传统建筑模型的参考也同样重要。巨大的柱廊排列得整齐划一，使人脑海中浮现出许多伟大的建筑，有大卫•切波菲尔德设计的位于德国马尔巴赫的现代文学博物馆，还有灵感主要来自于上世纪不朽的理性主义传统、由乔治•拉西设计的位于意大利基耶蒂的学生公寓。

sible to recognize the great lessons of Adolf Loos and of Bauhaus. The result is not an isolated exploit, but an expression of a deep reflection on composition via pure volumes which may be found in the entirety of the firm's work. The purity of volumes as a means of generating spaces via the penetration of light establishes a continuity between exterior and interior that is inspired, in a less abstract and more complex form, by the experiences of Mies van der Rohe, without abdicating, at the same time, an emotional sense of intimacy.

The play with light and shadow, with surfaces and material textures and with the views to the surrounding countryside, nullifies the need for any sort of decoration, in line with the evergreen Loosian aphorism, "Ornament ist Verbrechen" – ornament is crime. Every volume, and every void carved into it, is an expression of a precise functional role in the whole composition, based more on relationships than on single elements. The result is a building that could easily be situated in the Twenties of the XXth century, in the Stuttgart of the Weissenhof.

Chestnut Tree Twin Houses/Luzern, Switzerland_Lussi+Halter Partner AG

To regard the imposing, geometrical facade of the Chestnut Tree Twin Houses can generate a strong sense of déjà vu, reinforced by the national origin of the architects.

It is impossible not to follow memory to the elementary and immanent form of Le Corbusier's Ville Savoye in Poissy, to its enigmatic presence that gives order to the world. However, Ville Savoye is merely a starting point, a formal, intellectual matrix in which to set the program of the twin residences. As in Ville Savoye, the volumes inscribed in the squared, regular perimeter are independent from it, generating voids and internal views. As in Le Corbusier's villa, the entire building can be traversed using ramps. The terrace in which Le Corbusier places his "objects à reaction poetique" to allow the enjoyment in privacy of the surrounding countryside, here, with similar intentions, hosts a sculptured swimming pool. By the way, the introduction of the wall that separates the two houses, winking from the great cut on the principal facade, is sufficient to delineate a symmetry that would hardly have been found in Le Corbusier. Not less definitive of the departure from the model of reference is the use of the materials of the outside and interiors.

Chestnut Tree Twin Houses brilliantly demonstrate the possibility of reconsidering without reverential fear a monumental theme of a great master, suiting it to the circumstances and to the present time.

Sainsbury Laboratory/University of Cambridge, UK_Stanton Williams Architects

Many fathers can be credited with the Sainsbury Laboratory, in both the scientific and architectural fields.

The laboratory is set in the University of Cambridge's Botanic Garden, created by Regius Professor of Botany Reverend Professor John Stevens Henslow, mentor and friend of Charles Darwin.

References no less important can be discerned in the architectural profile. The monumentality of the giant order of the great colonnade

废墟上的希腊石柱
Greek columns in the ruin site

实际上,这幢大楼的精神遗产超越了其不朽的外观。其工作场所设计的最大特点就是在屋顶设置自然采光系统,这让人想起勒•柯布西耶在威尼斯设计的一家医院和伦佐•皮亚诺设计的梅尼尔收藏博物馆。另外,有些人也许早已注意到,很难不在以下两条通道之间建立理想的联系:侧面全是巨大的玻璃窗并把各个实验室连为一体的室内通道,以及查尔斯•达尔文自己住处Down House里他非常喜欢并称之为"思考之路"的通道。

然而,除了几个具体的、相对近期的设计参考,公共空间这一由来已久的设计理念也让人们感到非常熟悉。这一理念贯穿整个建筑史,从古希腊的拱廊和古罗马城市广场开始,传承到中世纪建有柱廊的城市街道。

Riberas de Loiola的耶稣教堂/圣塞巴斯蒂安,西班牙_ 拉斐尔•莫尼奥

拉斐尔•莫尼奥本身就是一位优秀的教师,但当面对一个新的宗教设计主题时,他没有忘记勒•柯布西耶提倡的基本建筑理念,追求更加兼收并蓄的方法。这一方法在其洛杉矶圣母大教堂的设计中得到了应用。

正如勒•柯布西耶设计的拉图雷特修道院,洛杉矶圣母大教堂的建筑设计语言可以说应用在了一个非常基础的体量之中,只增加了教堂用于进行基督教礼拜仪式的典型要素——钟楼、门廊、唱诗班、管风琴、十字架——尽管其灵感明显来自于中世纪十字架的象征意义并忠于宗教建筑的传统作用,但都根据简约、当代的建筑语言进行了重新设计。

遵循古哥特式教堂的传统,内部空间的垂直比例和从屋顶上方穿透的光线都能够传递强烈的精神情感。

即使在教堂的低层附设一个超市这一不寻常的做法,也许看起来不伦不类、稀奇古怪,但却可以追溯到中世纪宗教社区经济自给自足的传统,体现了宗教社区不仅拥有依靠他们的精神权威把周围的人聚集在一起的能力。

海德堡城堡游客中心/海德堡,德国_马克思•杜德乐

海德堡城堡游客中心项目表现的是一种类比关系,可归纳为最基本的以下三点:建筑材料、设计基本原理和建筑语言。

材料所传递的信息远超过形式本身。这里,原始的砂岩让人们想起了中世纪和文艺复兴时期军事建筑的力量。原始岩石材料的使用意味着厚厚的墙和凹入式的窗户,但是石头的组合使用体现了一种永恒的品味,超越了当代建筑。没有丝毫的媚俗或迎合本地审美的模仿,这一新建筑与著名的海德堡城堡废墟优雅共存,琴瑟和鸣。

作为海德堡城堡建成四个世纪后第一位获得允许在其区域建造

recalls to the memory many projects, from the Museum of Modern Literature in Marbach am Neckar by David Chipperfield to the Student's House in Chieti by Giorgio Grassi, which for their part have drawn largely on the tradition of the monumental rationalism of the last century.

Actually, the spiritual heritage of this building goes beyond its monumental facade. The workplaces are strongly characterized by a system of natural illumination through the roof that reminds one of Le Corbusier's hospital in Venice or Renzo Piano's Menil Collection. Furthermore, as others have already observed, it is difficult not to establish an ideal relationship among the internal path flanked by the huge glass panes connecting the laboratories, and the "thinking path" so loved by Charles Darwin in his Down House. However, more relevant than a few specific and relatively recent references, there is in the building a general feeling of familiarity with a sedimented idea of public spaces that cuts across the entire history of architecture, beginning from the ancient Greek stoa and the Roman forum, and passing on to the colonnaded streets of the medieval cities.

Iesu Church in Riberas de Loiola / San Sebastian, Spain_Rafael Moneo

Although himself a great teacher, Rafael Moneo does not forget the basic lessons of Le Corbusier when confronted with a new religious theme, after the more eclectic approach introduced in Our Lady of the Angels Church in Los Angeles.

As in Le Corbusier's church of the monastery of La Tourette, the architectural vocabulary is here reduced to an elementary volume to which are added only the archetypal components of a building devoted to the Christian cult – the bell tower, the pronao, the choir, the organ, the cross – all revisited according to a minimalistic and contemporary language, even though decidedly inspired by the medieval symbolism of the cross and true to the traditional program of religious architecture.

In continuity with the tradition of the ancient Gothic churches, the vertical proportions of the inside space and the light filtering from above are capable of transmitting an intensely spiritual emotion.

Even the unusual introduction of a supermarket in the building's lower level, which may seem atypical and quirky, has its roots in the medieval tradition of the economic self-sufficiency of the religious communities and in the ability to aggregate people around them not merely on the basis of their spiritual authority.

Heidelberg Castle Visitor Center / Heidelberg, Germany _ Max Dudler

This project represents an analogy reduced to the most basic terms of material, of grammar, of language.

A material can communicate more than the form itself. Here the raw sandstone evokes the power of the military architecture of the Middle Ages and the Renaissance. Raw stone means thick walls and recessed windows, though the stone is here composed in timeless, more than contemporary, taste. With no hint of kitsch or vernacular imitation, the new building coexists in elegant resonance with the famous ruins of Heidelberg Castle.

注释：
1. Friedrich Wilhelm Nietzsche, *Nachgelassene Fragmente Frühjahr–Herbst* 1881.NF-1881,11[202]
2. Quatremère de Quincy, A. C., *Dictionnaire d'Architecture, Encyclopédie méthodique*, Paris, 1832
3. Viollet Le Duc, *Dictionnaire raisonné de l'architecture française du XI au XVI siècle*, Paris 1867-1873
4. Giorgio Grassi, *La costruzione logica dell'architettura*, Padova, 1987
5. Bruno Taut, *Modern Architecture*, London; New York, 1929

圣玛丽修道院，法国Eveux-sur-l' Arbresle, 勒•柯布西耶, 1960年
Couvent St. Marie de La Tourette, Eveux-sur-l'Arbresle, France, by Le Corbusier, 1960

新建筑的建筑师，马克思•杜德乐抓住了海德堡城堡历史和遗址的精髓，使游客中心与原先幸存下来的建筑完美地交融在一起，也使新老建筑自然而然地形成了对照。

游客中心设计的历史灵感也扩展应用到周围的空间，应用到石头铺成的人行漫道上。建筑内部静谧而纯洁，通过精确安装的凹入式窗户，更加完美呈现了这一历史名胜迷人的景色。

Moka住宅/马德里，西班牙_ A-Cero

并不是所有现代建筑的特点都是形式和功能严格地一致对应。例如，加泰罗尼亚现代主义和有机主义都是新的反古典主义表述，但代表了不同的建筑方法。尽管许多建筑大师，如安东尼•高迪、路易斯•多明尼克•依•摩塔内尔、弗兰克•劳埃德•赖特、保罗•索拉尼，均深受当代技术所带来的新机遇的影响，但是他们发展了富有创见的关于形状、构造、体量和材料的建筑理念，其灵感或来自于植物和动物世界，或来自于得到极大改进的历史传统。这种方法带来了折中的、曲线优美的、依稀呈现生物外形的形状和体量，其设计感性多于理性。

尽管随后几代建筑师没有发现与上述建筑的做法有多少共通之处，但应用这一建筑方法的实例还是屡见不鲜的。毫无疑问，如果不是出自一些建筑大师之手，这种方法应用起来要冒更大的风险，总是容易被误认为是装饰主义。

从Moka住宅流畅的垂直线条中我们也能发现路易斯•巴拉干或保罗•波多盖西的设计思想，设计态度传承自这些最优秀的巴洛克风格建筑师，但根据严格而精益的简约主义进行了改进。

卡斯卡伊斯城堡酒店/ Cascais,葡萄牙_ Gonçalo Byrne + João Alexandre Góis + David Sinclair

通过一个类似于大脑中镜像神经元的同化过程，这个建筑模仿了城堡壁垒的开垛口设计。当然，这绝对不是纯粹的模仿行为，而是将新建筑本身融入历史遗迹中，是对历史遗迹的尊重和赞美。

与古城墙、大海迷人的景色和酒店简约而奢华的内部空间构成惊人的对比，这一建筑项目大胆并出色地重点解决了如下问题：如何改变这一充满回忆的地方？如何使这一充满回忆的地方重新焕发光彩？旧建筑和新建筑完美地交织在一起，没有让人有任何的搪塞和不和谐之感。

在这里，允许建造这样一个建筑物的有关当局功不可没；他们理解这个建筑项目的战略价值以及伟大的建筑师是出色完成本项目的保障，这位建筑师应该触觉非常敏锐，尊重环境和历史。这一新建筑没有丝毫敬而生畏之态，提升了这一历史遗址的品质，因此成为一个主要的旅游景点，造福了当地经济。

Max Dudler, the first architect allowed to add a building in the Heidelberg Castle area after four centuries, captured the very essence of the history and the site. The visitor center blends itself perfectly among the surviving original buildings, and naturally accepts comparisons with them.

The historical inspiration extends also to the surrounding space, to the stone-paved promenade. The silent purity of the interior perfectly enhances, through the recessed and precisely located windows, the intriguing views of the historical context.

Moka House/Madrid, Spain _ A-Cero

Not all of modern experience is characterized by a rigid correspondence between form and function. Catalan Modernism and Organicism, for instance, represented distinct approaches to a new anticlassical language. Although deeply involved in the new opportunities offered by contemporary technology, the great masters of these tendencies – Antoni Gaudí or Lluis Domenéch i Montanèr, Frank Lloyd Wright or Paolo Soleri – developed a visionary repertoire of shapes, textures, volumes and materials, inspired either by the vegetal and animal world, or by a largely revised historical tradition. This approach resulted in eclectic, curvaceous, vaguely biological shapes and volumes which were more emotional than rational.

Although these lessons have found less resonance in subsequent generations, there have been examples of this approach, undoubtedly more risky and always at the edge of decorativism, when not governed by the sure hands of the masters.

In the sleek vertical lines of the Moka House are also found the lessons of Luis Barragán or Paolo Portoghesi, with an attitude remotely inherited from the best baroque architects, but revised in the light of a rigorous, lean minimalism.

Cascais Citadel Hotel/Cascais, Portugal _ Gonçalo Byrne+João Alexandre Góis+David Sinclair

Via a process similar to the mirror neuron's assimilation in the brain, the building mimics the crenellation of the rampart. Definitely not a mere imitation, but an act of respect towards and admiration for the historical site into which the new building wedges itself.

In astounding contrast with the ancient walls, the enchanting view of the ocean and the minimalistic, luxurious interior spaces of the hotel, the project boldly and brilliantly addresses the problematic issues stemming from altering and revitalizing an area so rich in memory. Old and new perfectly blend themselves with no reciprocal prevarication.

The merits of the authorities who have allowed such an operation must not be overlooked; they understood the strategic value of the project and the guarantee offered by a great architect who is very sensitive and respectful of environment and history. The new building enhances the qualities of the site with no hint of reverential fear, and thus constitutes a major attraction for tourism and a boon for the economy of the community. *Aldo Vanini*

卡斯卡伊斯城堡酒店

Gonçalo Byrne + João Alexandre Góis + David Sinclair

卡斯卡伊斯城堡位于里斯本附近，是特茹河上一座重要的军事战略防御工事，它的规模、体量和室内形态决定了其建于城市网络形成的初期。

西特德勒（Citadel）城堡的历史由几个阶段构成，但都作为防御工事。

在近四个世纪的时间里，这个世代流传下来的城堡经历了许多变化，丧失了其用途，致使其败落。卡斯卡伊斯城堡酒店项目开始要修复的几个区域和建筑的情况都是如此。

卡斯卡伊斯城堡建筑具有定义城市的所有城市特征元素，因此，卡斯卡伊斯城堡酒店建筑项目为其提供了一个对这一历史遗产进行翻新/修复的极难得的机会，为人们梦寐以求的城市再生创造基本条件。

建筑师以谨慎的循环利用设计，改变了卡斯卡伊斯城堡原来的军事用途。通过这种用途的改变和建筑结构的修复来达到城市再生的目的，并使其空间和建筑结构适应民用与城市旅游业的发展，凸显卡斯卡伊斯作为旅游目的地城市的角色。

因此，卡斯卡伊斯城堡酒店项目一是要增加城堡世代传承的价值，二是要进行两个层面的修复改动。

首先，项目要对现存的有重大历史意义的建筑保持其供人欣赏的特性，一是可以对其用途循环利用/恢复其用途，或是对其进行修复，重新定位转型其空间，提升改进公共空间。

其次，就是新建筑的建设，在不打乱已有建筑类型的情况下，拟建新的建筑类型。通过使用轻型结构和当代建筑材料，新的建筑类型要寻求可逆性的价值观和令人称道的完整性。

卡斯卡伊斯城堡与旧时所有城堡要塞的结构形式和形态特征都是一样的，有一个中央广场，四个错落有致的结构单元环绕四周。

位于被称为Praça de Armas南面的建筑是整个项目的枢纽，the Port of Arms（通往城堡内部的唯一一条主要道路）保证了其轴心的地位，酒店位于广场和城堡区域的中心位置，那儿有入口和接待中心，位于一个"看见和被看见"的位置，这对酒店的服务功能来说是非常重要的，也确保了其接待场所非常宽敞。

在接待中心旁边的建筑一楼有一些可以看到广场的房间，接待中心从这栋建筑开始向南延伸至一个宽敞的配送空间和宴乐空间。

这个新建筑的地点以前是军队餐厅所在地，起到了从结构上把接待中心和酒店其他地方连为一体的作用。

建筑物外表采用了单一的毫无其他装饰的耐候钢筋网格，这样做是为了自动消除给周围建筑带来的影响，为了稳定/规定现有建筑物的价格。

在South Battery那边新建了一栋建筑，可以提供一些新的客房。

这一新建筑位于现存的要塞建筑的上部，强调透明度，与保持要塞雄伟庞大形象的墙体形成鲜明对照。

这样的话，这一新建筑形成了延伸的姿态，外观颜色清新淡雅，形成檐口阴影，淡化了庞大的城堡过渡到天空的突兀感。

项目名称：Cascais Citadel Hotel
地点：Cascais, Portugal
建筑师：Gonçalo Byrne, David Sinclair, João Alexandre Góis
合作方：G.B. Arquitectos_Hugo Guerreiro, Paulo Street, Joana Quintas Monteiro, Bruno Marcelino, Bernardo Bessa, Tomás Bonifásio, Carla Vieira / DSAA Arquitectos_Simon Dillon, Tiago Rocha, Tânia Cortez Pinto, Gonçalo Duarte, David Carvalho
装修设计：Jaime Morais, João David
结构工程师：Abel Almeida, Paulo Marinho
电气设计：Manuel Maçana
卫浴设计：Manuel Resende
景观设计师：Luisa Estadão, Gerald Luckhurst
气候调控设计：Raul Bessa
室内设计：Jaime Morais
甲方：Grupo Pestana SGPS, SA
施工方：Soares da Costa Grupo SGPS, SA
施工管理：Ana Melo
用地面积：17,500m²
建筑面积：8,000m²
总建筑面积：12,000m²
设计时间：2008
竣工时间：2012
摄影师：©Joao Morgado(courtesy of the architect)

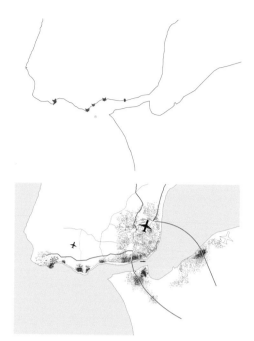

military defensive structure
introverted space
closed to the environment

historic hotel and culture space
extroverted space
the search of the environment

概念示意图 concept diagram

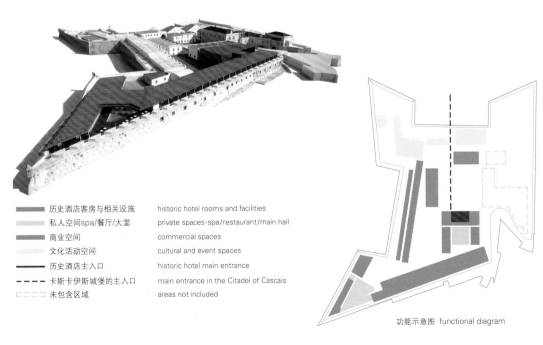

历史酒店客房与相关设施　historic hotel rooms and facilities
私人空间spa/餐厅/大堂　private spaces-spa/restaurant/main hall
商业空间　commercial spaces
文化活动空间　cultural and event spaces
历史酒店主入口　historic hotel main entrance
卡斯卡伊斯城堡的主入口　main entrance in the Citadel of Cascais
未包含区域　areas not included

功能示意图 functional diagram

Cascais Citadel Hotel

The Citadel of Cascais, a key element in the military strategy of the defense bar of the Tejo River in the vicinity of Lisbon, constitutes a fortified set, whose dimensions, content and interior morphology determines an embryo of urban network and approaches it of a stretch of an introvert Development City.

The history of the Citadel is made of phases that obey its own rules of a defensive structure.

The changes that this patrimonial set suffered along its nearly four centuries of existence resulted in a loss of uses, and resulting degradation, which was clear in several areas and buildings which are in the beginning of this project.

Structured with all the urban characterization elements definers of city, the Citadel of Cascais clothes itself thus of a unique opportunity for renewal/rehabilitation of the edified historical heritage, creating the basic conditions for a desired urban regeneration.

The conversion operation of uses and architectural rehabilitation seeks this regeneration by means of a careful recycling interven-

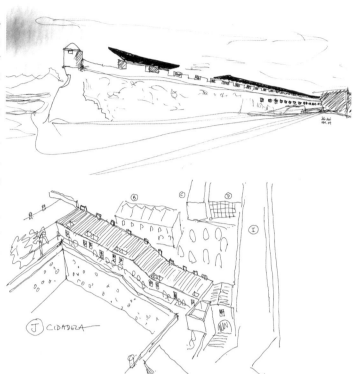

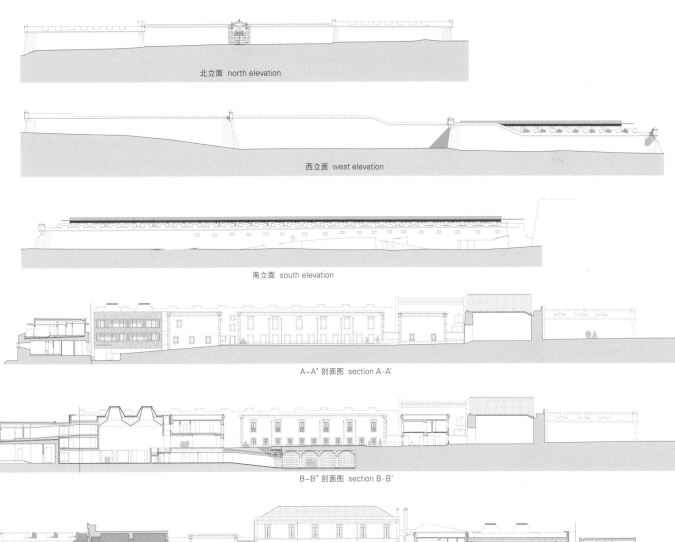

北立面 north elevation

西立面 west elevation

南立面 south elevation

A-A' 剖面图 section A-A'

B-B' 剖面图 section B-B'

C-C' 剖面图 section C-C'

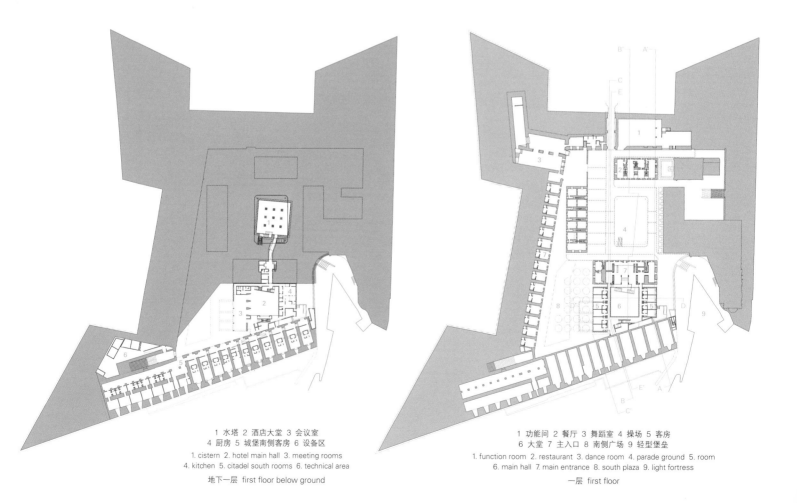

1 水塔 2 酒店大堂 3 会议室
4 厨房 5 城堡南侧客房 6 设备区

1. cistern 2. hotel main hall 3. meeting rooms
4. kitchen 5. citadel south rooms 6. technical area

地下一层 first floor below ground

1 功能间 2 餐厅 3 舞蹈室 4 操场 5 客房
6 大堂 7 主入口 8 南侧广场 9 轻型堡垒

1. function room 2. restaurant 3. dance room 4. parade ground 5. room
6. main hall 7. main entrance 8. south plaza 9. light fortress

一层 first floor

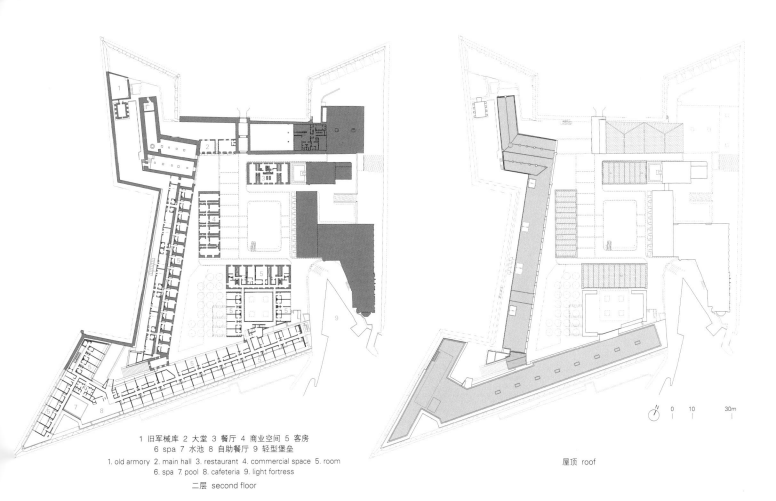

1 旧军械库 2 大堂 3 餐厅 4 商业空间 5 客房
6 spa 7 水池 8 自助餐厅 9 轻型堡垒
1. old armory 2. main hall 3. restaurant 4. commercial space 5. room
6. spa 7. pool 8. cafeteria 9. light fortress

二层 second floor

屋顶 roof

tion, amending its original military use and adapts the spatial and architectural support for a civilian use, urban tourism that highlights the role of Cascais as a tourist destination.

The action unfolds, therefore, on an intention of adding a patrimonial value and with two levels of action.

The first one with an appreciation character of the significant historic buildings existent, either by recycling/regeneration of uses, whether by its rehabilitation, requalifying spaces and promoting public space.

The second one, which translates into the construction of new buildings, that without violating the pre-existing typologies, proposes new typologies which seek the values of reversibility and recommended integrity, by using lightweight structures and contemporary materialities.

With organizational and morphological characteristics common to all the strongholds of the time, the Citadel of Cascais structures itself through a central square defined by four "buildings-block" properly hierarchized.

The building that is positioned in the south of the called Praça de Armas is assumed as the intervention hinge, allowing from the axis which is guaranteed by the Port of Arms (the main and only access to the interior of the Citadel), centering the Hotel in the Area of the Square and the Citadel, by locating there its Entry and Reception ensuring a position to "see and be seen", fundamental to the features that are proposed and to assure the character of a wide reception area.

From this building, which besides the reception, contains some rooms on the first floor overlooking the square, the reception stretches south to the large distribution space and conviviality, an inner courtyard with triple right foot.

This new construction, established in the same area previously occupied by the military canteen, building that without any equity value, establishes the organizational link between the reception and the rest of the hostel.

This building coated in corten steel grid, establishes a unitary and austere materiality, always with the aim of self-cancellation, always valorizing the existing buildings.

Over the South Battery is implanted a new body of rooms.

This new body is perched over the existing fortified structure and materializes with a great character of transparency, as opposed to the wall itself, opposition that allows the maintenance of its massive volumetric image.

This way, the new body arises stretched, with a subtle coating that allows establishing a cornice shadow that guarantees the dilute of the volumetry in its transition to the sky.

Gonçalo Byrne + João Alexandre Góis + David Sinclair

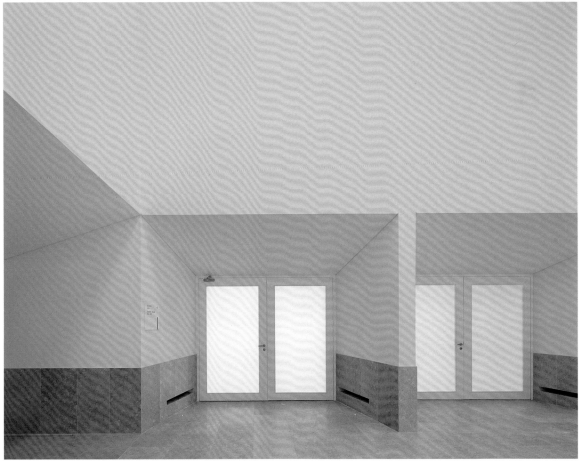

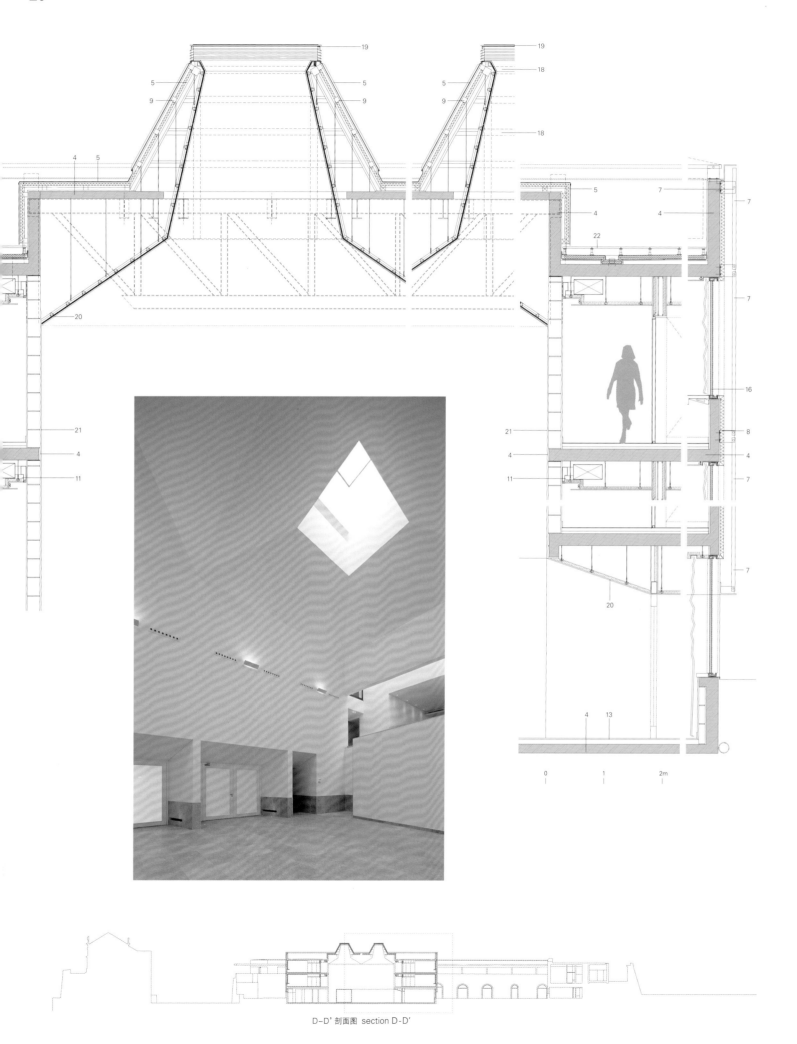

D-D'剖面图 section D-D'

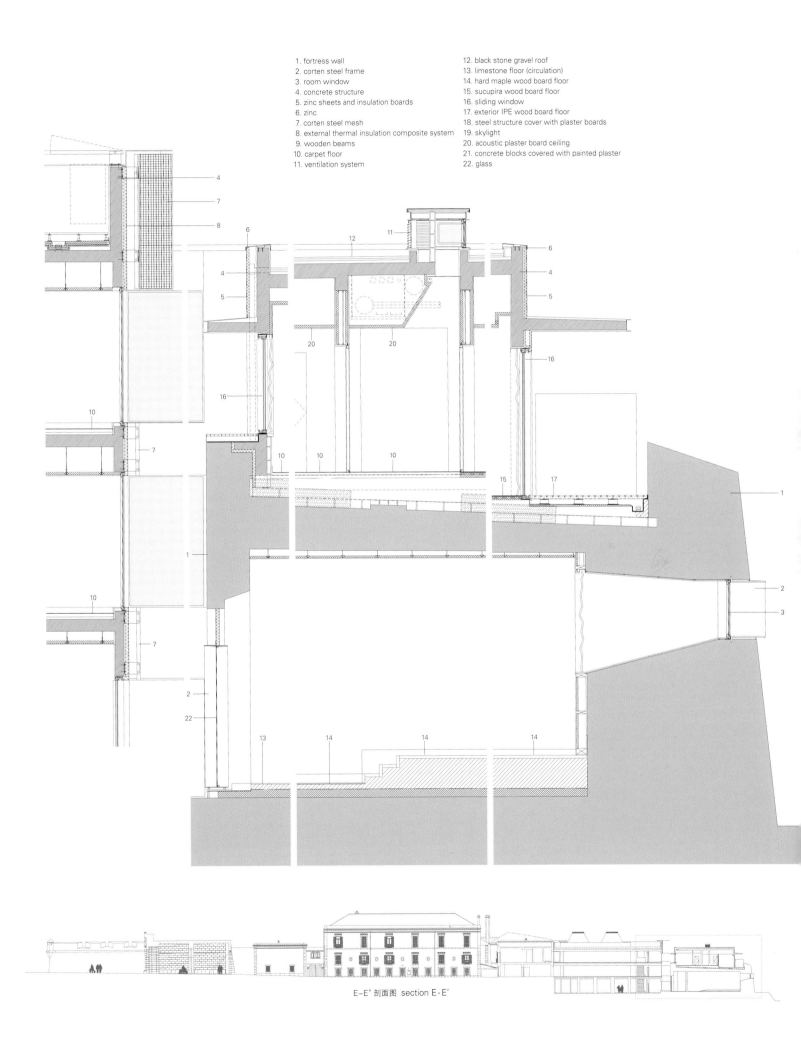

E-E' 剖面图 section E-E'

海德堡城堡游客中心

Max Dudler

海德堡城堡被列为阿尔卑斯山以北最重要的文艺复兴时期建筑之一。城堡在"三十年战争"期间已经部分毁坏，随后又经历了许多劫难，18世纪时城堡就被彻底废弃了。

如今，这一著名的废墟已成为一座博物馆。博物馆每年接待一百多万人次的游客，成为德国最受欢迎的旅游景点之一，给到德国游览的世界各地游客留下了久久难忘的印象。

建游客中心的目的是帮助游客在游览城堡区之前熟悉城堡。游客中心展示城堡的历史，引导游客，使他们在游览城堡时能够毫无障碍。2009年5月，马克思·杜德乐的设计在竞标过程中技压群芳。2010年夏天，游客中心奠基，成为四百多年来被允许在海德堡城堡建造的第一座新建筑。这一建筑向人们展示了马克思·杜德乐的当代建筑风格如何根植于历史的沃土。与此同时，建筑抽象的形式凸显了这一德国文化历史遗迹的宏伟和真实。

新建筑坐落于城堡旧的防御环墙以外，在通往城堡和宫殿花园的入口处。用来建造新建筑的那块狭长的地段位于一个小花房和一个建于弗雷德里克五世统治时期的马具店之间，背靠一段17世纪保留下来的、用来加固公园露台的挡土墙。游客中心的建筑线与周围建筑保持一致，其雕刻般的设计在结构上与城堡前院区域的其他建筑构成了一套完美的组合。

从建筑的角度来说，游客中心通过对场地上已有建筑的建筑元素的重新阐释，完全融入到了周围历史悠久的防御工事中。例如，窗户嵌进墙里2m多深，与旁边马具店大尺寸的开口相呼应。游客中心窗户安放的位置与建筑内部的要求相一致，也为游客从视觉上提供了与城堡外部入口建筑和花园的紧密关系。尤其是著名的伊丽莎白门，更是在建筑物内部的许多位置都可以看到。建筑的特殊布局使其正面深嵌的窗户成为可能：宽阔的外墙上隐藏了许多小房间和一个楼梯井，就像衣服的口袋一样，这些内部凹槽可以用来放置陈列柜、货架和座位休息区，而这一狭窄建筑的中心仍然是开放的。

游客中心的立面由机器切割后的当地内卡河谷砂岩构成，是一面经粗略切割的巨大的石砌块墙，几乎看不见石块与石块之间的接缝。这一砌石细部设计用当代的方式对历史悠久的挡土墙进行了重新演绎——这面墙是由手工切割的、未经任何加工的石材砌成的。不同于建筑物外部如浮雕般凹凸不平，建筑内表面非常光滑。大大的玻璃窗与白色的墙壁形成鲜明的对比，与此呼应，照明面板也内嵌入白色的天花板中。地板由淡蓝色的抛光水磨石铺成。凹槽处所有的固定装置和配件以及门和其他家具都是樱桃木做的。

游客中心建筑设计所面临的一个特殊挑战就是要让大量游客有序畅通地走动，杜德乐设计了一条贯穿整个建筑的"漫步长廊"，巧妙地解决了这一难题。游客从入口大厅进入教育展示室，然后上楼到屋顶平台，从高处远眺城堡，再经由游客中心后部的室外楼梯出去，开始城堡之旅。这样，就实现了这座小型建筑的全部潜力，既保证其具有多种功能，又保证了最大的游客通过量。

Heidelberg Castle Visitor Center

Heidelberg Castle ranks as one of the most important Renaissance buildings north of the Alps. Having been partially destroyed during the Thirty Years' War, and on many occasions since, the castle was abandoned altogether in the eighteenth century.

Today the famous ruin serves as a museum. Receiving more than one million visitors a year, it is one of the country's top tourist destinations and makes a lasting impression on international tourists visiting Germany.

The purpose of the visitor center is to familiarize guests with the castle before they proceed to the castle proper. The visitor center showcases the castle's history as well as orientating guests so as to ensure a trouble-free visit. In May 2009, Max Dudler's design prevailed in the architectural selection procedure. The visitor center's foundation stone was laid in summer, 2010 making it the first new building to be constructed at Heidelberg Castle for more than four hundred years. This building shows how the contemporary architecture of Max Dudler is rooted in history. At the same time, its abstract form underscores both the grandeur and actuality of this German cultural monument.

The new building is situated outside the old defensive ring wall, at the entrance gate to the castle and garden (Hortus Palatinus). The narrow strip of land chosen for the new structure lies between a small garden house and a saddle store built in the reign of Frederick V. The building backs onto a seventeenth century retaining wall which shores up the park terraces above. With its building lines following those of its neighbors, the sculpturally designed visitor center structurally completes this small ensemble of buildings in the forecourt area.

In architectural terms, the building blends in with the surrounding historical fortifications through its re-interpretation of elements of the existing site's architecture. The window embrasures, for ex-

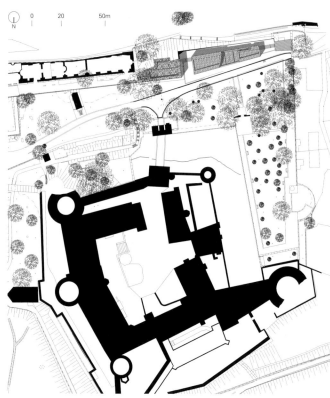

ample, are set more than two meters into its walls, echoing the large-sized apertures that can be seen in the neighboring saddle store. The windows of the visitor center are positioned according to the building's interior requirements and also offer visitors new visual relationships with the entry building and garden outside. The popular Elisabeth Gate in particular can be seen from many parts of the interior. The facade's deeply-set embrasures are made possible because of the special layout of the building: the broad expanse of its exterior walls hide a number of small side rooms and a stairwell. Like pockets (French: poches), these interior recesses offer space for display cabinets, shelves and seating areas, while the center of the narrow building remains open.

For the facade, local Neckar Valley sandstone has been machine-cut to form a monolithic wall of roughly-cut blocks with joins that are barely visible. This masonry detailing is a contemporary re-interpretation of the historical retaining wall, with its hand-cut, undressed stonework. Unlike the heavy relief of the building's exterior, the surfaces of its interior are smooth. The large window panes are fitted flush with the white plastered walls, as are the lighting panels set into the white plastered ceilings. The floor consists of a light blue polished terrazzo. All the fixtures and fittings in the recesses, as well as the doors and other furnishings are made of cherry wood.

Ensuring a smooth flow of large numbers of visitors was a particular challenge posed by the architectural brief. Dudler's design solves this with its ingenious "architectural promenade" through the building: visitors proceed from the entry hall through to the educational room, then up onto the roof terrace with its elevated views of the castle before exiting via the exterior stairs at the rear of the building to begin a tour of the castle proper. In this way, the full potential of this small building is realized, ensuring it has both multi-purpose usage and allows the maximum throughput of visitors. Max Dudler

项目名称：Heidelberg Castle Visitor Center
地点：Heidelberger Schloss, Schlosshof 1, D-69117 Heidelberg
建筑师：Max Dudler
项目经理：Simone Boldrin
合作商：Patrick Gründel, Julia Werner
结构工程师：Ingenieurburo Schenck
施工监理：plan-art
建筑物理及声学：ITA Ingenieurgesellschaft für technische Akustik mbH
建筑设备：IFG Ingenieurgesellschaft für Gebäudetechnik
景观设计：TDB Landschaftsarchitektur GbR
甲方：Land Baden-Württemberg (Vermögen und Bau Baden-Württemberg Mannheim Office)
建筑面积：490m²
总建筑面积：770m²
设计时间：2009
竣工时间：2011
造价：EUR 3,000,000
摄影师：©Stefan Müller(courtesy of the architect)

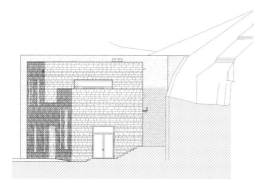

西立面 west elevation

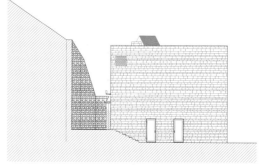

东立面 east elevation

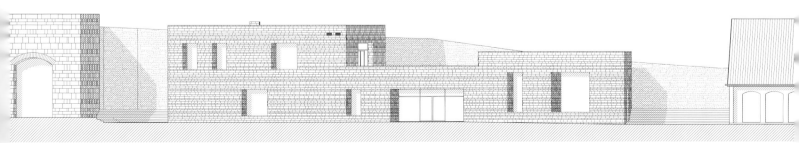

北立面 north elevation

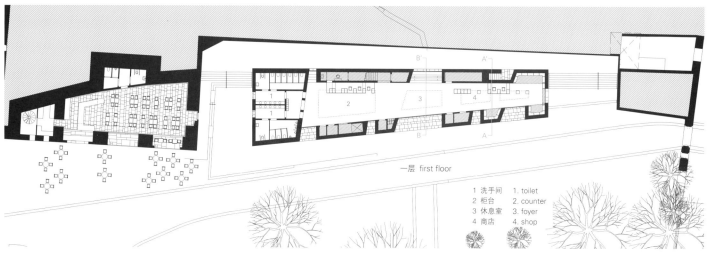

一层 first floor

1 洗手间	1. toilet
2 柜台	2. counter
3 休息室	3. foyer
4 商店	4. shop

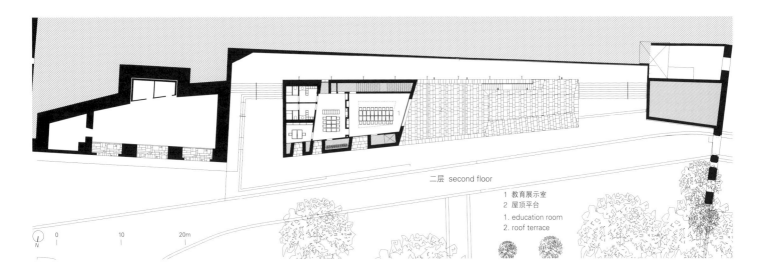

二层 second floor

1 教育展示室
2 屋顶平台

1. education room
2. roof terrace

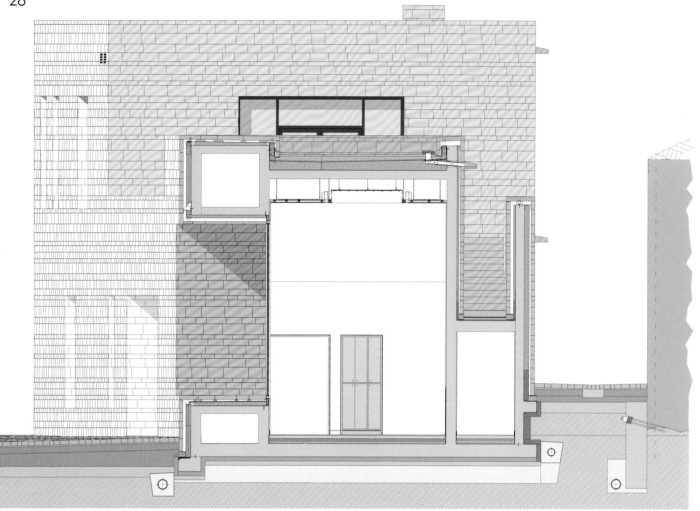

A-A' 剖面图 section A-A'

B-B' 剖面图 section B-B'

Moka住宅
A-Cero

项目名称：Moka House
地点：Pozuelo de Alarcón, Madrid, Spain
建筑师：A-cero
施工方：Vipecón Galicia s.l.
用地面积：3,035m²
建筑面积：1,520m²
总建筑面积：1,500m²
竣工时间：2011.12
摄影师：©Luis H. Segovia(courtesy of the architect)

Moka住宅位于马德里市郊，设计理念是从分布、布局和外观方面均能满足客户的需求。建筑设计为三层，力求使住宅需求得以完美实现。这栋住宅依一个一般坡度的地块而建，改变住宅公共区域内的主要房间，从而更好地利用周围风景及空闲的地面。

立面的建筑材料是混凝土。地下室既可以从住宅的主要区域进入，也可以从车库直接进入。

住宅有许多不同种类的区域：游泳池、运动设施、娱乐休闲区、维修工作区。从车库可以直接进入维修工作区，并且维修工作区通过楼梯与一楼相连；楼梯可通到经销商工作区。在中心区有一些储物架，那儿与车库相连，但没有直接照明和通风设备。

在一楼，通往住处的大厅通过一条东西向的虚轴连通两个不同的区域，一个是贵族区，另一个是对贵族区的补充，是穿过像卧室、起居室、餐厅和办公室这些主要房间的门阶。这条虚轴线的另一侧是一些辅助房间，如浴室、壁橱、餐具室和厨房。运动设施是一个室外的板球场，由住宅的主步行道与住处连为一体。

门廊的透空廊道使用轻型实心板建成，上面是钢结构。楼上为双层高度，从视觉上把上层和下层连为一体。双层高的楼上有卧室，专门供贵宾使用，相对于这个家庭的三个卧室而言，该卧室可以让贵宾享有更多的隐私，同时还与一间集图书馆、办公室和休息室为一体的房间相连。

这一正式的最终方案非常注重美观效果，是一个与本身高度、形状和周边环境和谐一致的纯粹空间。

Moka House

Moka House, located at the outskirts of Madrid is intended to respond to customer's needs in terms of distribution, placement and surfaces. It is developed in 3 levels. It has sought the perfect functioning in relation to the needs of the property which is adapting to an average gradient field and turning the main rooms of the communal areas of the development, looking for a better use of the views and use of free plot surface.

Concrete is the material of the facade. Access to basement, can be done either from the main area of the house or from the garage directly.

We can see many different kinds of areas; a swimming pool, sports facilities, place for relaxation and recreation, area of service work resolved by giving direct access from the garage and connecting it to the ground floor with a staircase landing in the dealer's work area. In the central area and connected to the garage are the storage stands, without direct lighting and ventilation.

On the first floor, the hall-dealer that provides access to housing, communicate two distinct areas by an imaginary axis running from east to west; one is the noble zone and the other is the area that complements the first one and gives way to the doorsteps through the main rooms as the bedroom, living room, dining room, and office. On the opposite side of this imaginary line, we find the additional rooms, bathrooms, closets, pantry and kitchen. The sports facilities are completed on the outside with a paddle court, linked to housing from the main pedestrian access section of the house.

The pergolas from the porch are made using light weight, solid slabs and on the top of these by metal structure with steel. The upper floor is visually connected with the lower floor by a double height in which you could find the bedrooms, reserved for distinguished guests, giving them more privacy, with respect to the 3 bedrooms belonging to the family and linked to a library-office and lounge.

The formal solution has been the result of an aesthetic look for categorical volume playing with their own heights and shapes and the surrounding environment. A-cero

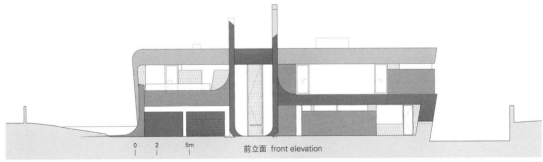

前立面 front elevation

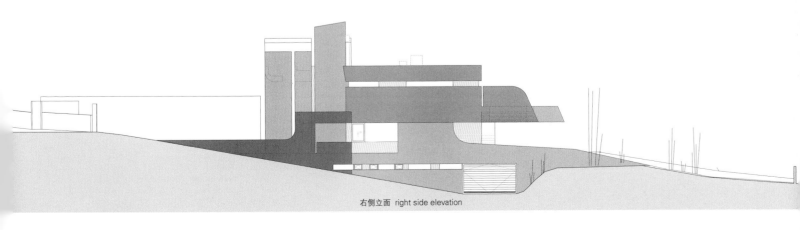

右侧立面 right side elevation

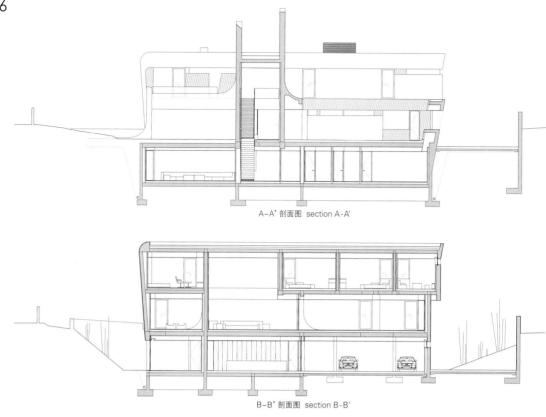

A-A' 剖面图 section A-A'

B-B' 剖面图 section B-B'

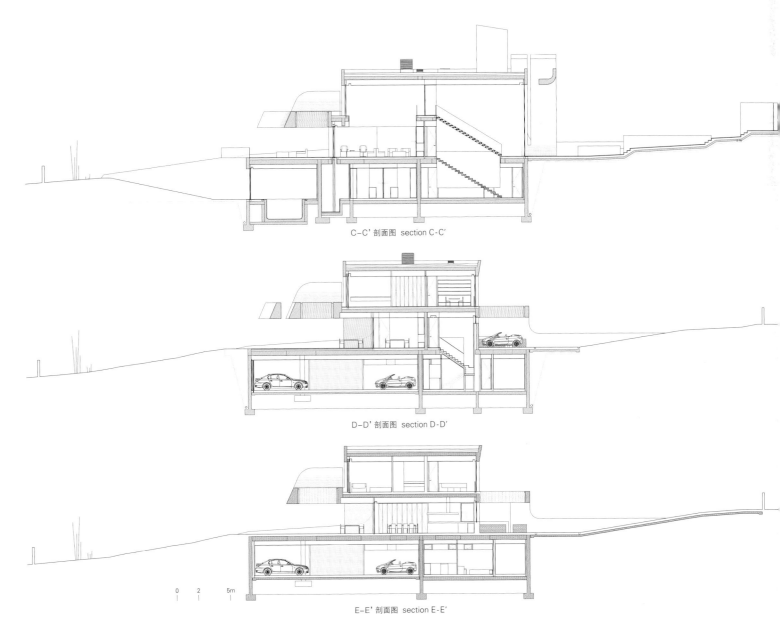

C-C' 剖面图 section C-C'

D-D' 剖面图 section D-D'

E-E' 剖面图 section E-E'

"代表"住宅
FORM / Kouichi Kimura Architects

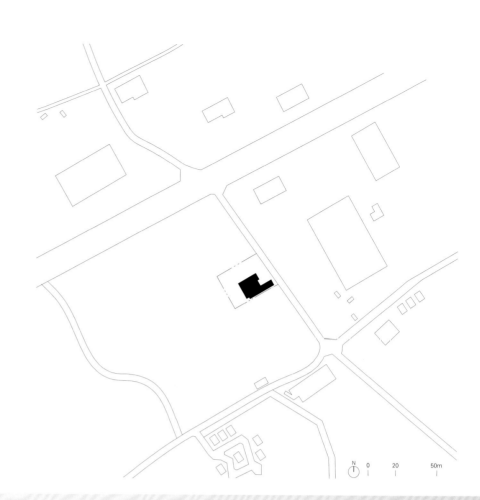

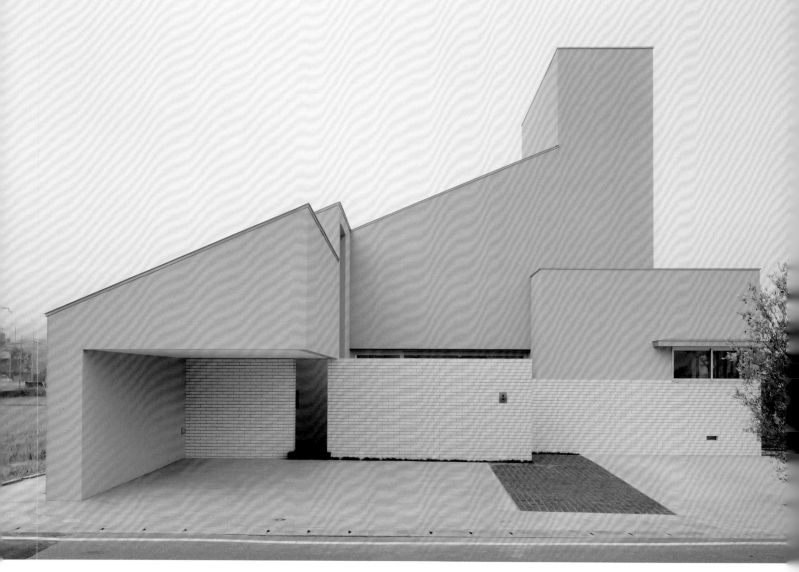

房屋位于一处可欣赏到乡村美景的地方。

客户要求房屋要给人以亲切感,但居住面积还要大一些。

房屋外部被设计成非常宏伟的形式,使其看起来像是对乡村景色的锦上添花。

整个房屋是一个向心的设计,一条贯穿多层起居区域的走廊将各个房间连为一体。

环绕走廊的墙壁,其形状和所用材质纹理都不一样,在空间内部形成一个序列。

自然而非直接照射的光线更加凸显了墙壁和各种空间,这样人们的目光就被内部装饰吸引,就会聚焦室内空间。

通过使用基本的对"光"和"景"加以操控的方法,设计师努力使内部空间显得更亲密、更宽敞。

设计师没有使用大手笔的、特技般的空间组合,而是通过一些控制光和观赏者眼球的、富有诗意的小变化创造了一个设计师本人认为非同寻常的空间。

东北立面 north-east elevation

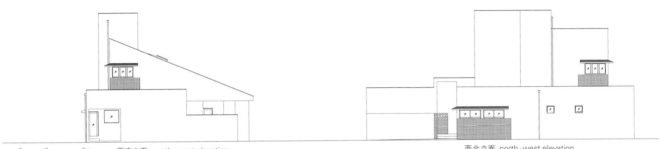

西南立面 south-west elevation

西北立面 north-west elevation

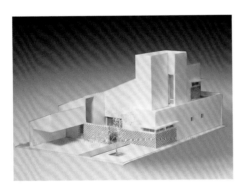

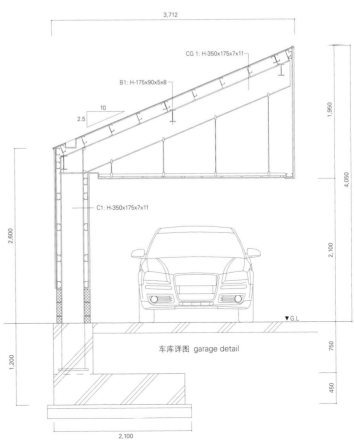

车库详图 garage detail

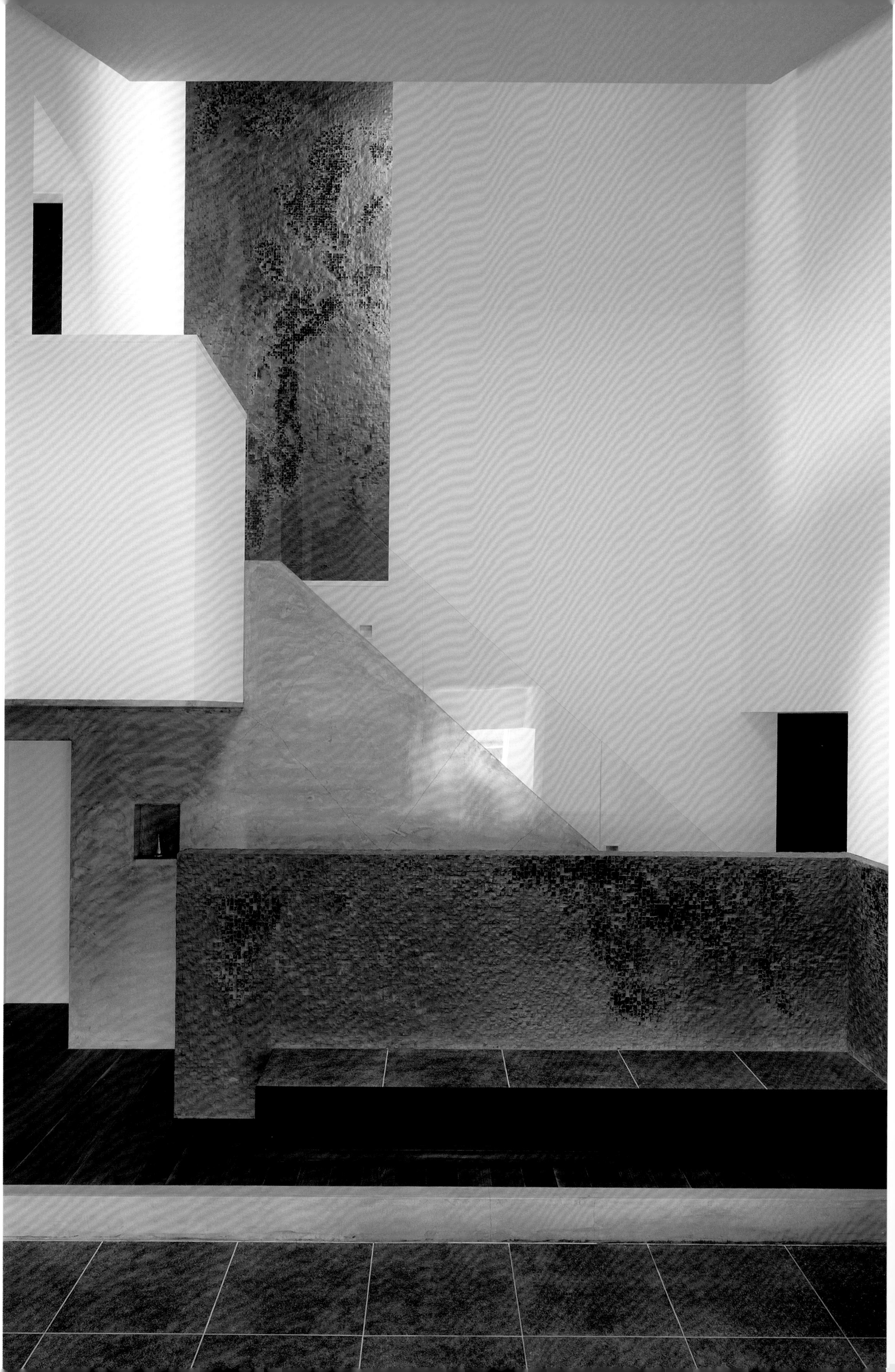

1	入口	1. entrance
2	客厅	2. living room
3	餐厅	3. dining room
4	厨房	4. kitchen
5	储藏室	5. storage
6	日式房间	6. Japanese-style room
7	儿童房	7. children's room
8	浴室	8. bathroom
9	庭院	9. courtyard
10	书房	10. study room
11	卧室	11. bedroom
12	阳台	12. balcony

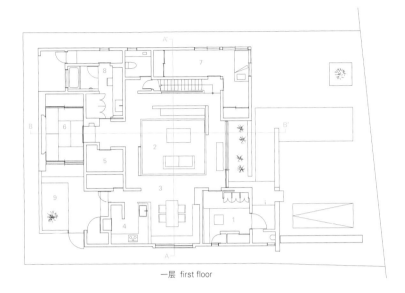

一层 first floor

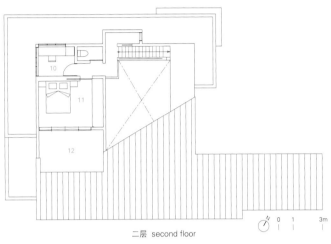

二层 second floor

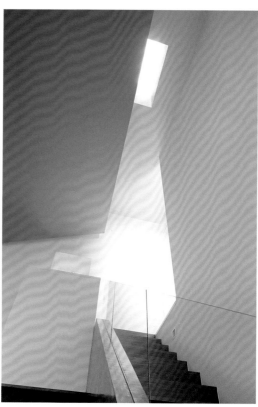

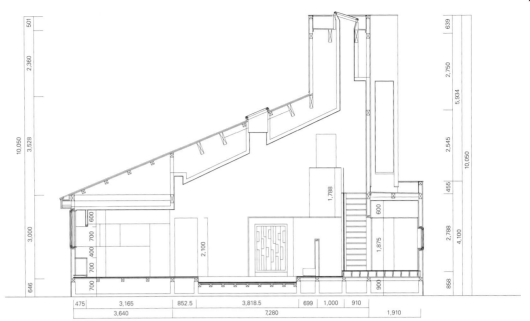

A-A' 剖面图 section A-A'

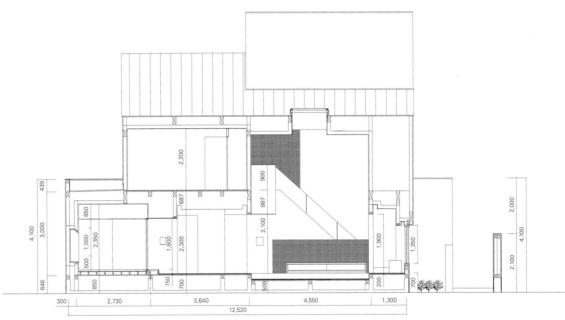

B-B' 剖面图 section B-B'

House of Representation

This house is planned on a site from which there are beautiful views of the country side.

The request from the client was for the creation of intimacy, but with a large living area.

The exterior is designed as a monumental form so that it can seem to be a new addition to the countryside scenery.

Architects created a centripetal plan where each room is connected through a corridor from the multi-level living area.

Around the corridor, walls, which have different textures and shapes, make a sequence inside the space.

The walls and spaces are highlighted by natural, indirect light, so your eyes are drawn to, and focus on, the interior.

By the basic processes of manipulating and controlling "light" and "views", architects tried to make the inside space more intimate and deeper.

Rather than doing large gestures of acrobatic space composition, architects have instead created what they feel is a space that represents the unusual, by doing small and poetic movements that control light and the viewers' eyes. FORM/Kouichi Kimura Architects

项目名称：House of Representation
地点：Kyoto, Japan
建筑师：FORM/Kouichi Kimura Architects
用地面积：355.78m²
建筑面积：213.20m²
竣工时间：2011
摄影师：©Takumi Ota(courtesy of the architect)

栗树双体住宅

Lussi+Halter Partner AG

湖与风景

这个独特的半独立式住宅位于卢塞恩湖滨的Kastanienbaum（"栗树"）社区，距卢塞恩市不远。实际上，房屋建在与湖毗邻的一块林地上，橡树、桦树、栗树和其他一些常青树木交错栽种。试想在林中漫步的感觉，光线透过浓密的树林，光影交错，共同把你引向一处鲜花盛开的草地。设计的目标正是留住这种感觉，从本案理念产生之初就有着重要的作用。本项目就位于这样一个风景如画的地方，建筑最小限度地占用了现有地面。门前先是混凝土露天平台，沿着坡道可以到达房屋，由混凝土建成的骑楼下面就是门廊。从那儿往前，宽宽的阶梯直通森林。新的室外空间努力保存原有的景色，浓密的树丛中满是桦树、栗树和矮树篱。

美丽的风景无处不在，增强了居住在大自然中的感觉。

内外的迷人对比

一楼一整面的玻璃外观使居住者尽享与周围环境直接接触的好处。建筑外围护结构看起来像是浮在空中一样，使其几乎完全融入到环境中，因此创造出一个明显的两用空间：只要调节木质卷帘，就可以把露台变成或是室内或是室外的一部分。这样，敞开的简易车库也能变成一个封闭的车库。颜色和材质的选择进一步加强了室内外之间的相互联系。黑色调的混凝土、微红色的南美柚木与大自然绿色的树影形成鲜明对照。从视觉上来说，深色材质装饰的室内空间并不显眼，会随着太阳光照在上面的强度和角度而变得绚丽夺目。镜面般的木制品表面把周围树木的光影反射到室内，更增强了整栋建筑与大自然的紧密联系。

作为连接元素的坡道

坡道贯穿整座建筑。从开放区域到私人空间，坡道作为连接元素把空间连为一体。清晨，来自东方的阳光穿过森林，照进一楼；下午，阳光从西面小山上照入上层楼面的室内庭院。这在视觉上延伸了上层楼面的居住空间，一段露天楼梯又把它与屋顶连接起来。坡道吸引着每个人像"散步"一样到房子里转转，同时给了居住在房子里的人开放、宽敞之感。

材料使用的极简主义

这栋双体住宅只使用了三种材料：一是黑色调的结构混凝土，用于墙体和天花板，而露台的地面是用坚固的无接缝混凝土建成的。其次，从巴西进口的非常坚硬的南美柚木，呈现出闪闪发光的微红色，用于铺设所有的室内地板、屋顶露台的木轨条以及从一楼能看到的露台的底面面板。因为南美柚木品质优异，它甚至可被用作浴室的淋浴地板。最后，设计师采用熠熠发光的无烟煤色的木制品增强了上述两种材料的使用效果。无烟煤色木制品的反光效果使厨房和橱柜犹如融进薄薄的空气中没了踪影。三种材料的组合营造出了一种安逸的心境和独特的氛围。材料之间和谐"律动"，如梦如幻，绚丽夺目。

项目名称：Twin Houses Kastanienbaum
地点：St. Niklausenstrasse 75, CH-6047 Kastanienbaum LU, Switzerland
建筑师：Lussi+Halter Partner AG
首席建筑师：Remo Halter, Thomas Lussi
协调建筑师：Corina Kriener, Aldo Casanova
机构工程师：Gmeiner AG
水利工程师：Markus Stolz
电气工程师：Jules Häfliger AG
照明设计师：Remo Halter, iGuzzini Illuminazione Schweiz AG
景观设计师：Koepfli Partner GmbH
施工方：Schmid Bauunternehmung AG
甲方：Remo Halter
用地面积：1,546m²
建筑面积：375m² (with terraces)
总建筑面积：400m² (without basement)
施工时间：2010—2011
摄影师：©Leonardo Finotti

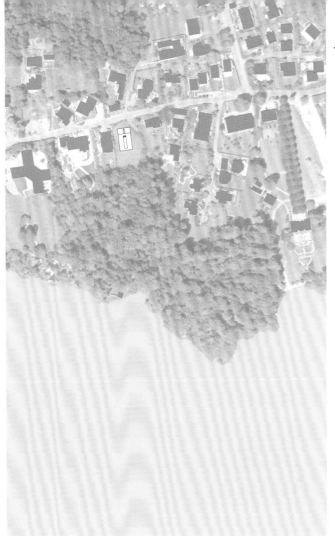

Chestnut Tree Twin Houses
Lake and Landscape

This unique semi-detached house is located in the community of Kastanienbaum ("Chestnut Tree"), not far from the City of Lucerne and on the shores of Lake Lucerne. In fact, the property abuts the lake at a woodland area mixed with oak, birch, chestnut and evergreen trees. Imagine the feeling of a stroll in the forest with its atmospheric interplay of light penetrating dense shadows, all leading to a flowering meadow; the goal of retaining this feeling played a major role from the project's very conception. The new construction is being located in this picturesque landscape in such a way that only a minimum of the existing terrain is being taken for the actual building. From a concrete deck, a ramp ushers you to the building where a concrete overhang defines the veranda area; from there, a wide stairway leads into the forest. The

1 游泳池
2 平台

1. swimming pool
2. terrace

屋顶 roof

1 宅地
2 有顶的阳台
3 主卧室
4 房间及工作室
5 浴室
6 客厅

1. curtilage
2. covered balcony
3. master bedroom
4. room & work room
5. bathroom
6. living room

二层 second floor

1 车库
2 走廊
3 入口
4 工作室
5 厨房
6 客厅

1. carport
2. veranda
3. entrance
4. work room
5. kitchen
6. living room

一层 first floor

1 园艺房
2 酒窖
3 桑拿房
4 淋浴室
5 洗衣房
6 娱乐室
7 浴室
8 设备间

1. room for gardening
2. wine cellar
3. sauna
4. shower
5. laundry
6. hobby room
7. bathroom
8. technical room

地下一层 first floor below ground

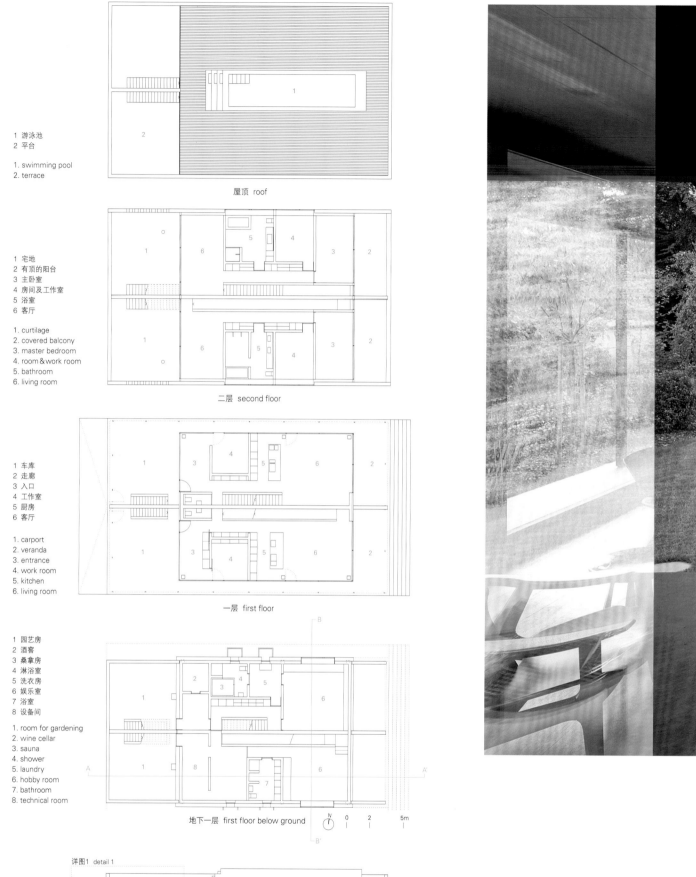

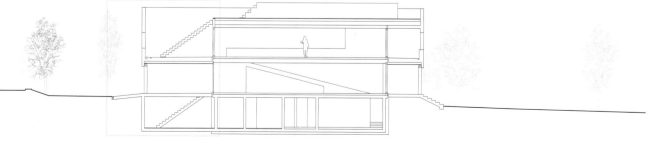

详图1 detail 1

A-A' 剖面图 section A-A'

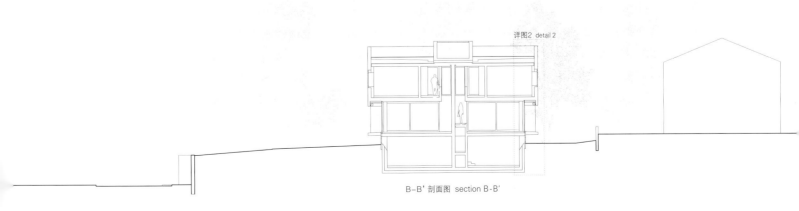

B-B' 剖面图 section B-B'

new exterior space attempts to preserve the existing atmosphere and is covered with a thick grove of birch and cherry trees as well as hedges.

The strong presence of the landscape reinforces the perception of living in nature.

A Captivating Contrast Between Inside and Out

A continuous glass facade on the first floor makes it possible for the inhabitants to enjoy a direct connection with their surroundings. Due to the building's seemingly floating envelope, it virtually merges into the environment and thereby creates a clearly identifiable dual-purpose space: depending on how the wooden roller shutters are adjusted, the terrace becomes part of the interior or outdoor space. In this way, the open car port is able to become an enclosed garage. The interrelationship between the interior and exterior is further enhanced by the selection of colors and materials. Black-toned concrete and reddish cherry wood (Jatoba) contrast to nature's shades of green. In optical terms, the interior spaces with their use of dark materials become inconspicuous, and the surroundings begin to glow depending on how the sunlight strikes them. The mirroring surfaces of the woodwork reflect images of the nearby trees into the interior spaces and further intensify the intimate bond the building has with nature.

Ramps as Connecting Elements

The length of the building can be traversed using ramps. These spatial interconnecting elements lead from open living areas to private quarters. The first floor receives morning sun from the east through the forest, and in the afternoon the sun shines from the west above the hills into the interior courtyard of the upper floor. This serves as an optical extension of the living spaces on the upper floor, and an outdoor stairway connects that floor to the roof level. The ramps invite everyone to experience the building as if "taking a stroll" and give those living there a feeling of openness and spaciousness.

Minimalism with Materials

In these twin houses, only three types of materials are used. First, the black-toned structural concrete is exposed at the walls and ceilings, while the terrace floors are made up of joint-free solid concrete. Next, the gleaming reddish and very hard cherry wood from Brazil (Jatoba) is used for all interior floors, the wooden railing on the roof terrace and the underside layers of the terraces as seen from the ground floor. Because this wood is of such high quality, this material can even be used as the shower floor in the bathroom. Finally, these first two materials are enhanced with built-in, gleaming anthracite-colored woodwork. Thanks to the reflections it throws off, the kitchen and cabinets seem to dissolve into thin air. This composition of materials alone creates a comfortable mood and unique atmosphere in this space. The materials "vibrate" in tune with each other and take on an incredible glow. Lussi+Halter Partner AG

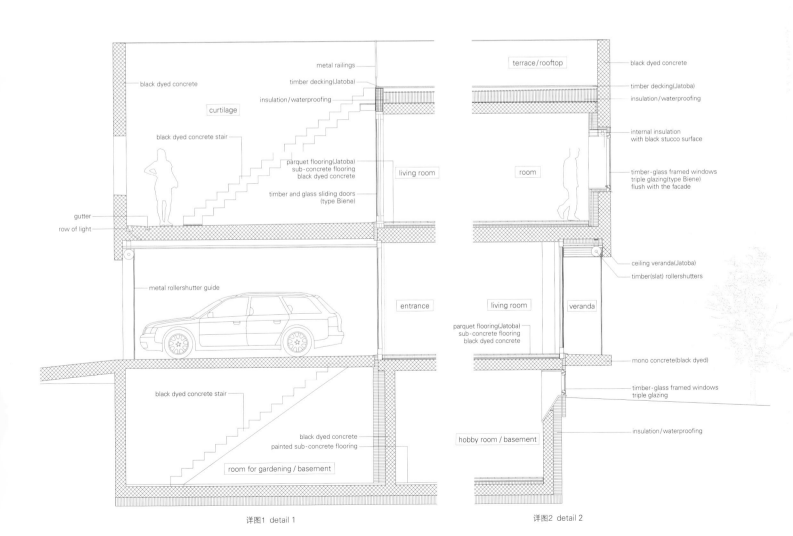

详图1 detail 1

详图2 detail 2

建筑谱系传承 Genealogical Reasoning

塞恩斯伯里实验室
Stanton Williams Architects

坐落在剑桥大学植物园的植物科学研究中心塞恩斯伯里实验室面积有11 000m²，这里汇集了国际一流的科学家，并为他们提供了最优质的工作环境。这一设计将复杂科学实验的高标准要求和建筑本身与环境相互协调的需要融合在了一起，提供了一个社区型的空间和激发研究的创新与合作的工作环境。建筑位于植物园中私人的"工作"区域内，包含了研究实验室以及相关辅助区域，还包括大学植物标本馆、会议室、礼堂、公共空间、员工的升级配套空间以及一个新的公共咖啡馆。

剑桥大学植物园由查尔斯·达尔文的良师益友约翰·斯蒂文斯·亨斯洛教授（英国著名的植物学家）在1831年构思设计，最初是作为植物多样性系统编目的研究工具而存在。塞恩斯伯里实验室发展了亨斯洛教授这一基本思想，寻求推进对植物多样性的产生理解。因此在设计上，对建筑与植物园进行整体考虑。建筑与植物园非常协调。

该建筑作为一个整体融入环境之中。地上可见建筑有两层，地下还有一层，这样做一方面为保证有效地环境控制，同时也是为了降低建筑高度，形成强烈的水平型建筑效果。建筑的坚固性体现在大量的石灰岩和裸露现浇混凝土的使用上，这会让人联想到地质岩层、达尔文的进化思想以及人们对这样一个重要的研究中心所期待的永恒性。但与此同时，建筑与植物园无论是实际的还是视觉上的通透性和连贯性也都是整座建筑设计理念的核心。

从建筑物的外观可以看出其身份。一系列连为一体而又别具一格、高度不一的建筑体占据着中心庭院的三面，中心庭院的另外一面生长着亨斯洛教授在19世纪种植的古老树木。建筑内部连通空间及公共区域围绕中心庭院布置，在一楼通向中心庭院，向上通向高抬的露台，这样，实验室与周围自然环境形成一种无缝连接。

对玻璃的精心利用进一步加强了建筑与周围景观视觉上的连接。在一层，大面积的玻璃使中心庭院和远处植物园的景色能够尽收眼底，将建筑内外有机连为一体。二楼也安装了巨大的玻璃。窗户被遮掩在狭长的石柱后面，石柱与玻璃窗一样高，未来可以根据需要改变开窗的方式。

建筑内部各区域由一条连续的路径连为一体，让人联想到达尔文的"思考之路"，这是通过步行来调和自然环境与人类思想的一种方式。从景观背景方面来说，这种处理方式与本建筑的设计理念紧密相关。此处，"思考之路"是大家思考和辩论的空间，目的是为了促进在此工作的科学家们之间以及他们与周围景色之间的接触、互动。"思考之路"一边是大面积的玻璃窗，可以看到中心庭院；另一边是室内的窗户，可以瞥见各个实验室，在位于建筑中心、开有天窗的工作区和植物园之间起到一个过渡区的作用。从这个方面来说，"思考之路"重新演绎了传统的希腊柱廊、修道院的回廊和学院的法庭，所有这些场所在某种程度上都成了人们沉思和集会的半室外空间。因此，塞恩斯伯里实验室建筑展示了过去、现在和未来的关联。实验室将为理解植物的多样性而不断探索，其所在的这座历史悠久的植物园正是以探索植物多样性而负有盛名的，它令到访者无比愉悦，并将继续作为一个致力于开创性研究的场所。

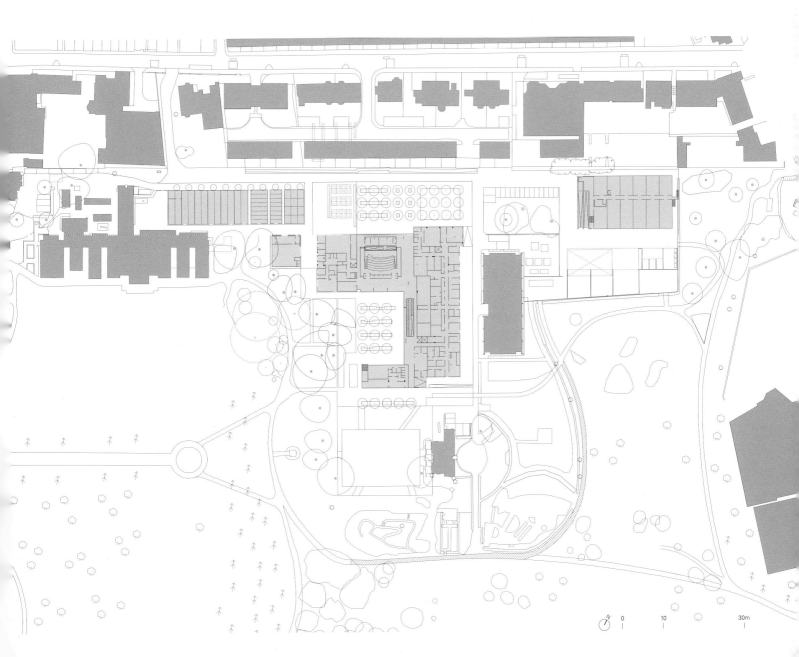

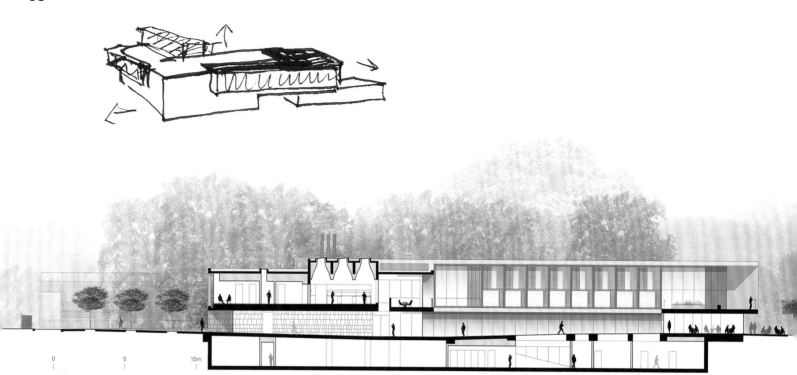

A-A' 剖面图 section A-A'

Sainsbury Laboratory

The Sainsbury Laboratory, an 11,000m² plant science research center set in the University of Cambridge's Botanic Garden, brings together world-leading scientists in a working environment of the highest quality. The design reconciles complex scientific requirements with the need for a piece of architecture that also responds to its landscape setting. It provides a collegial, stimulating environment for innovative research and collaboration. The building is situated within the private, "working" part of the Garden, and houses research laboratories and their associated support areas. It also contains the University's herbarium, meeting rooms, an auditorium, social spaces, and upgraded ancillary areas for Botanic Garden staff, plus a new public cafe.

Cambridge University Botanic Garden was conceived in 1831 by Charles Darwin's guide and mentor, professor Henslow, as a working research tool in which the diversity of plant species would be systematically ordered and catalogued. The Sainsbury Laboratory develops Henslow's agenda in seeking to advance understanding of how this diversity comes about. Its design was therefore shaped by the intention that the Laboratory's architecture would

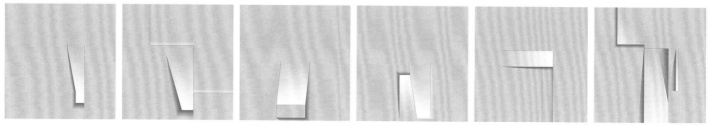

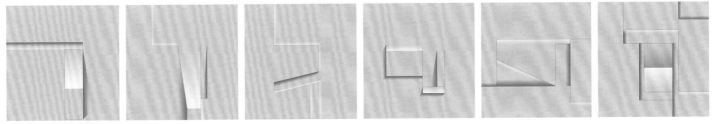

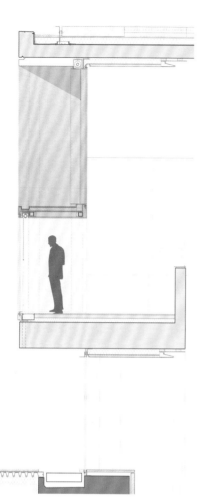

西南立面详图 south-west facade detail

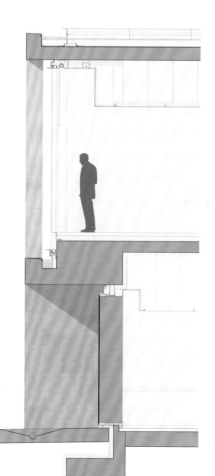
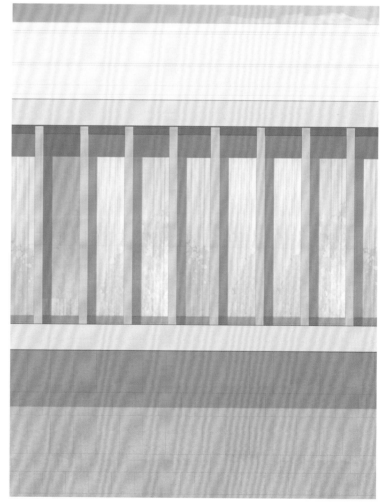

东北立面详图 north-east facade detail

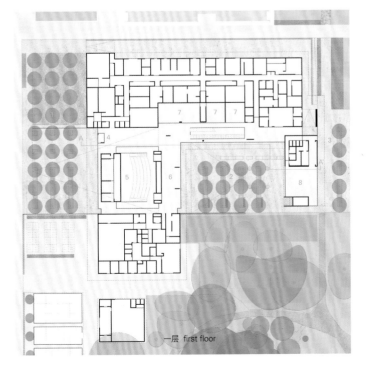

一层 first floor

二层 second floor

1 入口庭院	1. entry court
2 中心庭院	2. central court
3 咖啡厅平台	3. cafe terrace
4 主入口	4. main entrance
5 阶梯教室	5. lecture theater
6 员工餐厅	6. staff dining
7 会议室	7. meeting room
8 公共咖啡厅	8. public cafe
9 室外平台	9. external terrace
10 研究区域	10. study area
11 办公区域	11. office area
12 实验室	12. laboratory

express its integral relationship with the Garden beyond.

The building as a whole is rooted in its setting. There are two stories visible above ground and a further subterranean level, partly in order to ensure efficient environmental control, but also to reduce the height of the building. The overall effect is strongly horizontal as a result. Solidity is implied by the use of bands of limestone and exposed insitu concrete, recalling geological strata and indeed the Darwinian idea of evolution over time as well as the permanence which one might expect of a major research center. At the same time, however, permeability and connections – both real and visual – between the building and the Garden have been central to its conception.

The building's identity is established externally by the way in which it is expressed and experienced as a series of interlinked yet distinct volumes of differing height grouped around three sides of a central courtyard, the fourth side of which is made up of trees planted by Henslow in the nineteenth century. The internal circulation and communal areas focus upon this central court, opening into it at ground level and onto a raised terrace above in order to provide immediate physical connections between the Laboratory and its surroundings.

Further visual connections are created by the careful use of glazing in the building. At ground level, extensive windows provide views of the courtyard and the Garden beyond, allowing these internal areas to be read as integral elements of the outdoor landscape. The first floor is also largely glazed. Its windows are screened by narrow vertical bands of stone that imbue the elevation with a regular consistency, behind which the pattern of fenestration could potentially be altered in response to future requirements. Related to the conception of the building in terms of its landscape setting is the way that its internal areas are connected by a continuous route which recalls Darwin's "thinking path", a way to reconcile nature and thought through the activity of walking. Here the "thinking path" functions as a space for reflection and debate. It is intended to promote encounters and interaction among the scientists working in the building, and between them and the landscape. With glazed windows facing the court on one side and internal windows offering glimpses of the laboratories on the other, it operates as a transitional zone between the top-lit working areas at the center of the building and the Botanic Garden itself. In this respect, the "path" reinterprets the tradition of the Greek stoa, the monastic cloister, and the collegiate court, all of which were intended to some extent as semi-outdoor spaces for contemplation and meetings. As a result, past, present, and future are connected. The work of the laboratories will seek to understand the plant diversity that is glorified by the arrangement of the historic Botanic Garden in which it is set and which, though pleasant to visit, continues to function as a working space devoted to groundbreaking research. Stanton Williams Architects

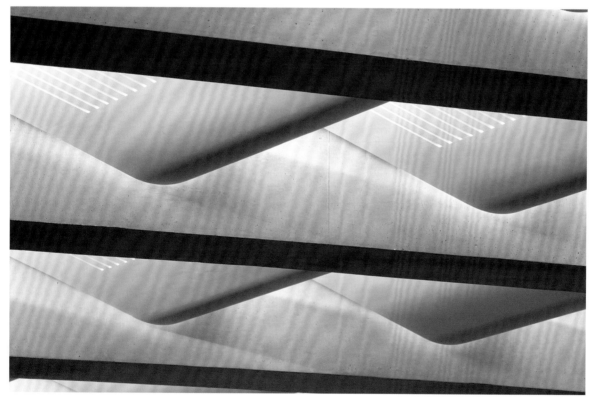

项目名称：Sainsbury Laboratory
地点：Bateman Street, Cambridge, UK
建筑师：Stanton Williams
项目经理：Stuart A. Johnson Consulting Ltd
土木及结构工程师：Adams Kara Taylor
建筑设备工程师：Arup
景观设计师：Christopher Bradley_Hole Landscape, Schoenaich Landscape Architects
CDM协调员：Hannah-Reed
承包商：Kier Regional
甲方：The University of Cambridge
总建筑面积：11,000m²
设计时间：2006—2008
施工时间：2008.2—2011.1
摄影师：©Hufton+Crow(courtesy of the architect)

建筑谱系传承 Genealogical Reasoning

Riberas de Loiola的耶稣教堂
Rafael Moneo

继洛杉矶大教堂之后，主教塞巴斯蒂安正在Riberas de Loiola的新居住区依蜿蜒曲折的乌尔玛河建一座教堂。在难得风平浪静的宗教建筑领域，这对建筑师来说是一次新的经历。如果说在洛杉矶，大教堂所需的象征内涵及其规模而让建筑师感觉胆怯的话，在Riberas de Loiola建的这座教堂则可以让建筑师把心里的这块大石头放下。这一次很明显，建教堂主要是希望在一个多元和多样化的社会中，为信奉基督教的群体提供一个活动空间。

成立教区是为了在新居住区的入口处为教民服务。新居住区通过公园与整个城市连为一体。可以看出，这座教堂绝对是一座占主导地位的建筑，因为教堂的尖塔会告诉人们这座教堂是一个轴心，社区内所有的公共生活都不可避免地围绕它转。这个教区紧凑而密集，我们经过时，以开放、友好的姿态展现在我们面前：教堂永远都向每一个人敞开怀抱，不仅仅是那些有相似想法和相同信仰的人。本着服务于人、贴近生活的愿望，教堂内设宗教法庭、大厅、教室、研讨室、住所，教堂低层设了超市，紧连着公园，不言而喻，这里是一个集会的地方：一个最新式的圆柱门廊下的开放市场。

进入教堂要经过像过滤器的空间，这个空间使我们意识到我们想要去教堂寻求什么。空间抽象而富有立体感，也许其垂直的体量促使我们将它视为一座卓越的宗教建筑，这也是哥特式教堂中令人欣赏的地方。十字架再次出现在屋顶，光透过屋顶照进教堂，这里再一次使用了隐喻的方式，引导我们将光线理解为一种超然的存在。

我想在结束这一简单的介绍之前提一下，所有参与这个教堂建筑项目的人都非常期待教堂有奥泰萨（西班牙雕塑家）和奇里达（西班牙艺术家）的作品，渴望在教堂建筑中发现他们的大师之作。

Iesu Church in Riberas de Loiola

After the Cathedral of Los Angeles, the church that Bishopric of Donastia-San Sebastian is building in the new neighborhood of Riberas de Loiola along the bend of the Urumea River is a new experience for the architect in the tempestuous territory of religious architecture. If in Los Angeles, the architect felt intimidated by the symbolic content needed for the Cathedral and by its scale, in the church in Riberas de Loiola, this weight was lifted. On this occasion it was clear that the expectation for the architecture was primarily to help configure a space capable of contributing to the life of the community that feels Christian in the midst of a plural and diverse society.

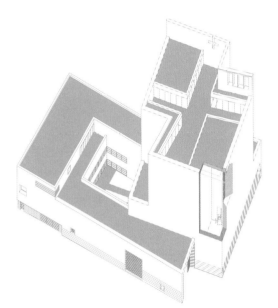

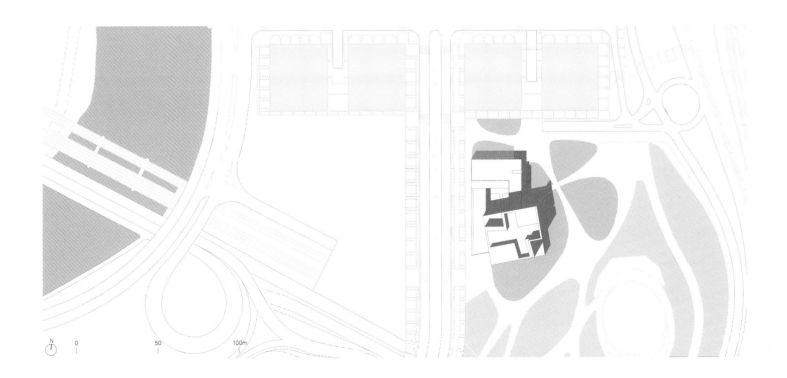

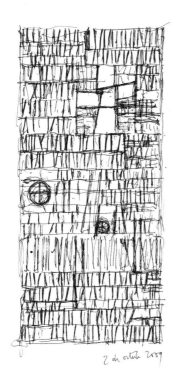

The parish came into being with a will to serve at the entrance to a new residential neighborhood that establishes continuity with the city through a park. The church is the dominant volume in the way that a steeple tower makes us understand the church as the axis around which all social life inevitably turns. The volume of the parish, compact and dense, presents itself as we pass by, offering itself to us with an open and friendly gesture: the church is a space always willing to receive everybody, not only those that share similar ideas and share the same faith. A court, a hall, classrooms, seminar rooms, residences and with the desire to serve and not be apart from the everyday, a supermarket in the lower level closely connected to the park, is understood as a meeting place: an up-to-date version of the open markets under in porticos.

The access to the church takes place through the filter of a space that makes us aware of what we look for when we go to church. The space is abstract, cubic and it could be that its vertical condition helps it to be seen as not far from what in our culture is understood as a religious space by excellence, the space we admire in Gothic cathedrals. The cross appears again in the roof, bringing in light, once again using the metaphor that leads us to understand light as a manifestation of transcendence.

I would not want to finish these brief lines without mentioning how much all of us involved in the construction of this church would have liked to have counted on the presence of the work of Oteiza and Chillida: the desire that a trace of their mastery can be found in the architecture of the church. Rafael Moneo

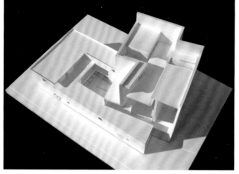 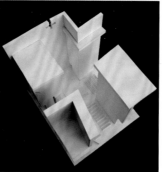

项目名称：Parroquia Iesu
地点：Avenida de Barcelona nº2, San Sebastián, Spain
建筑师：Rafael Moneo
项目团队：David Goodman, Eduardo Arilla, Pedro Elcuaz,
Isabel López, Luis Miguel Ramos, Fernando Iznaola, Gabriel Fernández-Abascal
结构工程师：Miguel Salaberría
设施：Estudios Tecnicos Eneka
项目发起人及承包商：Riberas del Urumea
甲方：Obispado San Sebastián
面积：church_1,300m², narthex_658m²,
parish center_1,704m², supermarket_2,244m², parking_4,720m²
造价：EUR 11 million
设计时间：2004
施工时间：2007—2011
摄影师：©Duccio Malagamba

A-A' 剖面图 section A-A'

1 社区活动室	1. community room
2 庭院	2. courtyard
3 教堂	3. church
4 洗礼堂	4. baptistery
5 小礼拜堂	5. chapel
6 公寓	6. apartments
7 超市	7. supermarket

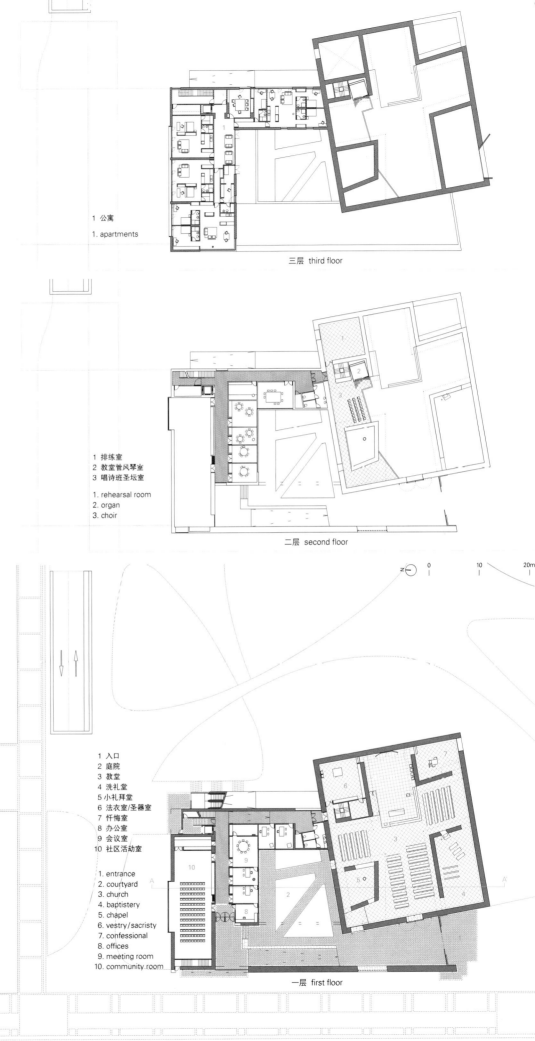

从限制到优势

建筑设计领域如果严格受限，新项目自由发挥的空间会有多大？在现有条件下，对新设计产生的严格限制多基于以下因素：地方性法规、文化文物保护条例、地理位置的特殊性（包括现存建筑的结构、样式和周围环境）。然而，即使受到这种限制，建筑师的自由发挥仍可将原先的限制条件转换成新设计中的关键因素，从而产生出人意料、令人惊讶的结果。新的解决方案可以巧妙地嵌入项目实施过程中，弥补已有限制所造成的缺憾，或者灵活地将其作为提高设计质量的起点。因此，建筑结构方面的限制能够被用来赋予一个地方与过去相联系的独特氛围。虽然地方性法规严格限定了建筑的外观，但建筑师仍有绝佳机会去精心设计建筑的内部，使室内外清晰的对比成为新设计的优势。

本节探讨了建筑受到严格限制和约束时所具有的表现力，提出了超前的解决方案。

How broad a range can a new project really span when the domain of the architect is restricted? Serious limitations for new design in a pre-existing condition can be set based on several factors, such as local regulations or cultural heritage and legacy stipulations, or certain conditions of the physical site, including existing structures, forms and contextual presences. Thus confined, however, the architect's freedom can result in unexpected and surprising results, with the original limitations becoming a key point in the new design. The new solution can snake around the impositions, trying to fill the gaps resulting from the restrictions, or it can deftly take them as a starting point for improving design quality. Structural constraints can thus be exploited in order to confer upon a place a unique atmosphere linked with the past. A rigid set of local rules imposing a fixed outward appearance of a building can be a good opportunity to create a sophisticated project on the inside, making the clear contrast a strong point of the new design.

This section investigates the expressive power of architecture when constrained and compelled, presenting unforeseen solutions.

Capelinhos火山讲解中心/Nuno Ribeiro Lopes Arquitectos
Fontinha码头/Alexandre Burmester Arquitectos Associados
圣伊莎贝尔住宅/Bak Gordon Arquitectos
安嫩代尔住宅/CO-AP
Ceschi住宅/Traverso-Vighy Architetti
Rizza住宅/Studio Inches Architettura
Zayas住宅/García Torrente Arquitectos
潜望镜式住宅/C+ Arquitectos
超越限制/Silvio Carta

Capelinhos Volcano Interpretation Center/Nuno Ribeiro Lopes Arquitectos
Fontinha Wharf/Alexandre Burmester Arquitectos Associados
Santa Isabel Houses/Bak Gordon Arquitectos
Annandale House/CO-AP
Ceschi House/Traverso-Vighy Architetti
Rizza House/Studio Inches Architettura
Zayas House/García Torrente Arquitectos
Periscope House/C+ Arquitectos
Over the Constraints/Silvio Carta

城市设计：超越限制
可恶的现实主义

在诸如限制、局限、约束等"阴影"的笼罩下，一种可能自由发挥的建筑设计理念会大打折扣。面对这样的前景，建筑师可能会不顾项目中建筑所处的实际地理位置，表达自己的设计理念。因此，我们将建筑师的构想投影到建筑实地上，这就像把在异地建好的建筑最终直接转移到实际地点，然而，这种对特定建筑项目的构想基于建筑师对目标设计的理解和想象（至少在最初阶段是这样），在一种自由和不受约束的氛围中——在设计过程的特定阶段——他们遇到了现实条件中"可恶的现实主义"。关于现代评论与理论这方面，建筑设计中的"可恶的现实主义"概念是在20世纪80年代后期由利恩•勒费夫儿提出的。[1]从那时起，这一概念被广泛采用并引起诸多争论。例如，勒默•凡•托恩[2]在他的文章中介绍了这一概念，认为其与塔伦蒂诺的《低俗小说》有相似之处（"塔伦蒂诺能使我们更敏锐地阅读现实主义吗？在《低俗小说》中，没有对声音和图像进行过多的电影蒙太奇手法的特技处理，但是小说中舞台调度所描绘的场景能使我们将日常生活中的现实理解为'可恶的现实主义'[3]"），而且他对库哈斯有关公共领域和文化的分析进行评价时也提到了这一概念（凡•托恩说"库尔哈斯使用的是令人讨厌的视觉文化"[4]）。

变化的观点

在应用"可恶的现实主义"这一概念时，我们默认建筑项目及其所处的环境是一对矛盾体。虽然设计活动体现了自由思想的人文表达，但涉及到"实际"问题时，"现实"又将这种高高在上的设计理念带回到较低层面。从这一角度而言，"现实"传递了特定的负面涵义，就本质而言，它与积极的设计活动相背离，因为设计活动被看作是具有创造性和建设性的做法。从这一层面而言，有些建筑师坚持认为设计师应该面对并处理好现实存在的"可恶的现实主义"，而另一些建筑师则将"真实条件"所带来的"限制"视为其设计存在中的挑战：开启其设计的有利条件。在20世纪80年代末和90年代初，第一类建筑师围绕"后批评主义"开展了大量的专题讨论，其中值得一提的有伯纳德•屈米、雷姆•库哈斯和弗雷德里克•詹姆逊。[5]在各种立场之争中，有一种观点认为，考虑到设计背景，建筑师的设计作品应该和普通大众的批评立场相分离。"后批评主义流派"[6]的观点对于处理客户的需求和设计背景条件这类复杂问题提供了解决办法。要处理建筑项目偶尔出现的一些令人抓狂的难题，或者用库哈斯的话来阐述，"要解

Urban How: Over the Constraints
Dirty Realism

The cloud of words such as *restriction, limit, constraint* entails a shrinkage imposed upon a supposedly free architectural idea. Faced with such a prospect, the architect expresses his or her design concept regardless of the reality of the actual place in which the project will eventually sit. Thus we have a projection of the vision of the architect onto the reality of the building plot, which appears to have been formulated in another place and eventually transferred directly onto the final site. However, this way of seeing certain architectural projects is based on the assumption that architects think about and design objects (at least in their preliminary phases) within a sort of free and limitless dimension and that – at a certain point in the design process – they encounter the "dirty realism" of actual facts. In relation to contemporary criticism and theory, the notion of "dirty realism" in architecture was introduced by Liane Lefaivre in the late eighties.[1] Since then, the concept has been widely employed and debated. Roemer van Toorn[2], for example, introduced that notion in his texts, via parallels with Tarantino's *Pulp Fiction* ("*Does Tarantino enable us to read reality more perceptively? In Pulp Fiction, it is not so much the filmic montage of sound and image, but the mise en scene that induces us to comprehend everyday reality as a 'dirty realism'*[3]") or in his assessment of Koolhaas' analysis of the public sphere and culture ("*Koolhaas uses the visual culture of pulp*"[4] says van Toorn).

Changing Perspective

In applying the notion of "dirty realism" we implicitly agree to describe architectural projects as being in a conflictive relationship with their (real) context. Whereas design activity represents human expression of freedom of thought, factual "reality" brings high ideas back to a lower level, where "real" things happen. To this extent the "reality" conveys certain negative connotations which are, by default, set against the positiveness of the design activity, which by contrast, is seen as a progressive and constructive practice. In this respect, while some architects continue to argue the existence of a "dirty reality" that designers should somehow face and work with, others consider the "constrainedness" imposed by "real conditions" as a challenging aspect of their design: an advantageous position to start from. For the first group, the seminal discussion sprang forth during the end of the eighties and early nineties, centering on the post-critical and featuring mentions of such people as Bernard Tschumi, Rem Koolhaas and Fredric Jameson.[5] Amongst the various possible positions, the idea emerged that the work of the architect should be detached from a general critical position with regard to the context of the

决经济、文化、政治和社会生活等方面积累产生的一些难以解决的问题，需要共同参与，如果要用一个世俗化的词语来替代，那就是'共谋'一词，而我很坦白地用它来代替共同参与或者达成一致这样的表述"。7

现有的立场

与此相反，第二类建筑师则积极应对现实条件中存在的限制，即使它们不是设计中内在的要素，他们也将其视为建造任务的一部分。可以用一个比喻来阐明这一立场。现实环境带来的这些"限制"（无论是地理位置方面的、世俗规范方面的或是文化方面的），都被当作是项目工地中已经矗立起的"丰碑"。乍一看，这些现有的"丰碑"似乎构成了问题所在，给建筑设计的自由带来潜在的限制，这些"丰碑"预先排除了项目可能会呈现的一系列的构造方式。尽管和这些"丰碑"相关的批评的（或后批评的）立场可能会引发有关它们存在的合理性、对它们的实际需要或背后所蕴含的政治和意识形态领域方面的争论，但因此也带来了设计立场之争的问题，建筑师（无论是作为个体还是作为职业范畴）应该站在这一立场（"我们应该把这些'丰碑'推倒吗？"）。对此问题的非批评性解决方法是认为"丰碑"的存在理所当然，根本不属于讨论主题的范畴。因为这些"丰碑"确实已经矗立在那儿，成为建筑场地（或说周围的环境）既定的条件，建筑师面临的主要问题就衍变成如何在设计中利用这些"丰碑"。

两种策略

在"丰碑"存在的前提下，其设计可能采取的立场范围可以归纳为基于项目在场地空间实际表现的两种策略。第一种策略是让项目以"丰碑"为中心展开，项目的最终结构源于现有限制条件和建筑师理想二者的结合。第二种策略是在项目定位时，直接超越"丰碑"（包括设计理念和建筑本身），即超越实施策略，这种叠加的方式超越了第一种策略小修小补的方式，是原有建筑结构的扩展，或说是场地价值的提升。在这两种实施策略中，强加的限制条件使得项目本身必须呈现出建设性特征，并且初始条件将会在项目的结果中体现，除此之外，还需考虑项目各方面里（包括形式、空间和审美等）限制条件的处理方式。

由于地方法规、文化关怀、文物保护和地理位置条件所导致的复杂的初始条件而形成了"丰碑"，本节中的项目列举了几种处理"丰碑"的办法。

design. The post-critical stance[6] was an answer to the complex problem of dealing with clients' needs and contextual conditions. To cope with the sometimes insane difficulty of an architectural project, or to put it in Koolhaas' words, *"to deal with the incredible accumulation of economic, cultural, political, and logistical issues,"* is an undertaking which *"requires an engagement for which we use a conventional word – complicity – but for which I am honest enough to substitute the word engagement or adhesion"*.[7]

The Existing Standings

By contrast, the second group encompasses those architects who positively regard the limiting facts of the real context and consider them to be part of the assignment, if not already intrinsic aspects of their design. An allegory can be applied to clarify this position. The "constraints" imposed by the context (whether these constraints are physical, normative or cultural) can be considered as posts already present in the plot on which the project will sit. At first glance, these existing posts can appear to be a problem, imposing potential limitations upon the preliminary freedom of design. The posts exclude in advance an array of possible configurations that the project may assume. While a critical (or post-critical) position with respect to the posts may entail a debate as to their presence, the actual need for them, or the political and ideological implications behind them, and hence raise questions concerning the position the architect (as an individual and as a professional category) should take in that regard ("Should we remove the poles?"), an a-critical approach to the problem takes the posts for granted and does not consider them a subject of discussion. Given that the posts are actually "there" as pre-existences of the site (say surrounding conditions) the architect's main question becomes how to cope with them.

The Two Strategies

The range of possible positions the design may take with regard to the presence of the posts can be summarized in two strategies based on a physical action performed in space. The first is for the project to actually snake around the posts, with the project's final configuration resulting from this dialogue between the constraints and the architect's intentions. The second consists in the project being positioned directly above the posts (whether figuratively or even physically). The superimposition – more than the tailoring operation of the first case – is an addition to previous configurations or values of the site. In both cases the imposed limitations will inform the project in a constructive manner, and the starting condition will be manifested in the result, in addition to the way the impositions have been dealt with in each of the

安嫩代尔住宅，建在现有墙体之间的狭长地块上。
Annandale House, built on a long and narrow plot between existing walls.

Fontinha码头——它是面向波尔图多罗河的仓库的混合型翻修工程，由亚历山大·伯梅斯特建筑事务所设计，展示了一种可行的解决方法。该项目谨慎而精确地融进原有的建筑结构，位于原有建筑内部或者修饰着它们的周边。人们或许可以真正品味出"剪裁式设计"在平面、剖面以及立面这三个维度是如何实施的。它最终形成了一个三维立体、新老构件交融的结果，每一种构件都与其他构件多方面融合。新构件是在"可用"空间内新设计的物化体现，从而得体地翻新了破旧的墙壁和建筑结构。建筑师成功建成了全新的建筑综合体，它体现了新的设计理念和现有条件的积极"融合"。

另一方面，Ceschi住宅则是一个受限更多的案例。现有的建筑——部分斜靠在12世纪维琴察城的古城墙上——只能让建筑师特拉韦尔索·威戈在几乎固定、完全受约的"空间"内工作。其约束条件具体体现在现存古城墙的结构骨架上，以及依附于古城墙的原有住宅的"空墙"。该介入建筑包括在古城墙内置入一种新型轻便的结构，即"一种骨架……确定新建筑的结构布局，并使新老融为一体"，该项目的建筑师解释道。通过在新老构件之间建立一种清晰的新型关系，原有问题得到了妥善解决。在此案例中，新的建筑结构（满足新的抗震和功能方面的要求等）显然脱离了原有石墙。

由C+建筑事务所设计的潜望镜式住宅出现在西班牙受保护的村庄——奥拓安普达村原有的石墙之间。建筑师们将保护条例中有关规定视为设计规则，采取积极主动的办法来应对。在原有建筑条件的限定下，充分考虑客户的"建筑意愿"，该项目得以推进。该住宅围绕着历史遗迹的原貌所体现出的约束条件，即空间的分布、环形道路、结构元素和支撑模式，进行设计，但这些设计从外部几乎看不出来。在这个项目中象征意义也发挥了至关重要的作用，而这种新的融合方式被认为非常适度、不会引起他人的注意。该住宅的外部保留了原有历史遗迹的风貌，而其内部则充分表达了设计的自由理念。初始的限制条件经过精心设计转变成一种"隐形策略"，即"通过不断思考和讨论使所有事物都体现在该住宅的内部"，建筑师如此解释道。

Rizza住宅则展示了另一个内部设计自由表达的案例。该项目由瓦卡罗的尹彻斯建筑事务所设计。严格的瑞士建筑装修规定给该项目的翻新工作带来了巨大挑战。按照规定，建筑主体外观的主要特征（开口、屋檐的高度、屋脊和斜屋顶）须保持原样，Rizza住宅这一新项目大部分是在其外部不变的条件下进行改造的。将现有约束条件和

project's aspects (formal, spatial, and aesthetic).

The projects in this section illustrate several ways to cope with the "posts" imposed by such complex starting situations as local regulations, cultural concerns, legacy matters, and physical conditions of the site.

Fontinha Wharf – a mixed-program renovation project of warehouses facing Oporto's River Douro designed by Alexandre Burmester Arquitectos Associados – illustrates one possible approach. The project is a careful and precise intervention within the existing fabric, physically tailored around and within the in-there buildings. One may appreciate how the "tailoring" has been performed at the floor plan level, as well as in the sections and elevations. The result is a three-dimensional integration of old and new parts, each engaged in a complex dialogue with the other. The new parts are the materialization of the new design within the "available" spaces, resulting in a respectful approach to renovating the dilapidated walls and structures. The architects succeeded in realizing a new complex which represents a positive "merging" of new design intentions and the existing condition.

On the other hand, the *Ceschi House* represents a more delimited case. The existing structure – partially leaning on the 12th century wall of the city of Vicenza – offers to architects Traverso Vighy the chance to work within an almost fixed "container". The constraining condition is here embodied in the skeleton of the ancient existing wall and the "empty walls" of the existing house attached to it. The intervention consists of the insertion of a new lightweight structure: a "skeleton... defining the distribution and establishing a collaborative relationship with the container", explain the architects. The default condition is tackled by establishing an openly explained new relationship between existing and new parts, where the new structure (meeting new anti-seismic and functional requirements and so on) is clearly detached from the stone walls.

The *Periscope House* by C+ Arquitectos unravels amid the existing stone walls of the protected village at the Alto Ampurdá in Spain. The architects have here taken an active approach by "taking the conservation regulations as rules of the game". The project has been developed between the existing condition of the previous building and the clients' "building desires". Barely noticeable from the outside, the holiday house is conceived around the constraints that the ruined complex presents, namely the distribution of spaces and circulation paths, structural elements and supporting patterns. Symbolic values play a crucial role in this project as well. The new intervention is seen as a modest, non-attention-seeking approach. The outer shell of the *Periscope House* is flattened by its

圣伊莎贝尔住宅，反映出之前半工业棚户区的构造。
Santa Isabel Houses, reflecting the previous configuration of semi-industrial sheds.

新的设计理念进行结合所带来的结果影响了房子的布局。新项目在原有农场的仓储建筑（一座四层的6m×6m的塔式建筑）中进行。一座本来为与农场相关的工作——谷仓、为农民和家畜预留的存储区域——而设计和建造的建筑，现在改造成了一个私人住宅项目。整座建筑充分利用了原有条件，与新的介入部分形成了鲜明对比。整个建筑内部光滑、干净、明快，与17世纪的小镇相比，这一氛围显得更加浓厚；打开任何一扇门或窗，这些特征都一览无余。在建筑物内部白色的表面上跳动的自然光，对在外部的人来说是一种意想不到的景象。因此，建筑物外观保持原封不动的要求，最终为该项目指引了一个全新的发展方向。

澳大利亚的安嫩代尔住宅阐释了它的产生背景。钢琴工厂的砖墙、阶梯地形和市中心的结构模式，再加上城市规划委员会的管制条例，充分体现了现有的限制条件，但同时，它们也为建筑师设计独特的住宅创造了绝佳机会。从外部来看，最初狭窄的建筑用地（更像现有的两堵墙壁之间的城市狭长地带）本该从一开始就建成灰暗狭小、令人不爽的房舍。但是，CO-AP建筑师事务所将这些原先的困难转化为机遇，将住宅沿着场地的长向进行布局，并重新考虑了房屋开放和闭合的表皮设计。这一"狭长的建筑用地"和庭院连续呼应，屋顶的水平开口可以照进更多的自然光。该项目最终成为一个开放式、光线充足、全新的建筑结构。

里斯本的圣伊莎贝尔住宅的设计理念源于城市有限的空间格局。这种密不透风的模式已被用于建筑本身，也启发了位于里斯本的比克戈登建筑师事务所进行项目设计的灵感。现有的设计"约束条件"源于现有城市结构所导致的复杂空间布局，正如建筑师所解释的，这一特点"深深根植于街道、广场和街区之中"。建筑物带覆层的的表皮和天井都源于先前的构造，并且承载了某城区的形式和空间痕迹，该城区之前以半工业棚户区为特色。

位于葡萄牙亚速尔群岛的Capelinhos火山讲解中心非常清晰地将追溯建筑和景观存在的年代的意图具体化。整个建筑旨在铭记、解释和证明Capelinhos灯塔区域在1957至1958年间火山爆发后所发生的巨大变化。新的项目以保留废墟、重新考虑改变后的景观为前提，将"现有"建筑与其历史遗迹成功地结合在一起。以前的定居点遗址对于新讲解中心的建立至关重要。废墟不仅是现有的限制条件，同时也提供了一个基本场景，使人们再次体验"火山爆发的过程，将内容丰

previous picture, while the freedom is expressed within. The limiting original conditions have been elaborated into a "disappearing strategy" where "everything is interior, built through reflections and activity", explain the architects.

Another example of inward expression is provided by the *Rizza House*, by Studio Inches Architettura in Vacallo. Restrictive Swiss building renovation rules have informed the project in a challenging way. With the building forced to keep its envelope practically unchanged in its main features (openings, height of the eaves, ridge and type of pitched roof), the new program of *Rizza House* has been developed mainly behind its outer shell. One consequence of this combination of imposed restrictions and new design affects the house's typology. The new program is housed in the former farm storage building (a four-stories 6×6 m towerhouse). A building conceived and realized for purposes related to farm work – a barn and storage area for farmers and beasts – now houses a private residential program. The house takes full advantage of its old condition by creating a significant contrast with the new intervention. The slick, clean and sharp atmosphere created within is enriched even further when compared with the seventeenth-century context of the town, which may be viewed through any opening. The natural light bouncing all over the white surfaces of the interior is an unexpected feature from the outsider's view. The requirement to keep the building envelope untouched has thus eventually opened unforeseen directions for the project's development.

Annandale House in Australia, speaks of the context it originates from. The presence of the piano factory's brick wall, the stepped terrain and the pattern of the urban fabric of the inner city, along with the brief council planning controls, have represented constraints, but they have, at the same time, been opportunities for the designers to create a "unique" new house. The narrow original plot (more like an urban pocket between existing walls) would have suggested a dark, small, uncomfortable house at the outset. However, CO-AP architects turned these original questions into opportunities by stretching the residential program along the site's length and reconsidering the opening and closing surfaces of the house. The "stripe" has a continuous relationship with the courtyard, and horizontal openings in the roof admit more natural light. The result is an open-plan, light-filled new configuration.

Santa Isabel Houses in Lisbon, originate from the spatial limitations of the urban pattern. The "impenetrable" patterns have been used both physically and as an inspiration by the Lisbon-based Bak Gordon Arquitectos to create the project's configuration. The design

地下的Capelinhos火山讲解中心，位于灯塔遗址附近。
Subterranean Capelinhos Volcano Interpretation Center, adjacent to the ruins of the light house.

富、充满科学和休闲特征的博物馆内容细化为三个不同的部分：火山爆发前、爆发中和爆发后"，建筑师这样解释道。

艾尔·阿尔贝肯区（西班牙格拉纳达市的历史街区之一）所传承的文化和历史遗迹阐释了加西亚·托伦特建筑事务所在重新设计Zayas住宅时遇到的主要约束条件。实际上，该项目要对建于16世纪的摩瑞斯克住宅进行翻新，这种住宅属于典型的阿拉伯摩尔人的传统庭院建筑。新项目已围绕现有的建筑结构、空间和细节进行了严格翻修，以期与新要求相结合。除受到过去的限制或约束之外，Zayas住宅实际是以原有的建筑设计为蓝本的。从这一意义而言，看似限制的条件可能最终会导致意想不到的结果，而且可能会在很大程度上改善先前被遗弃和被忽略的建筑设计。

1 在利恩·勒费夫儿一篇文章的简写本中，她提出了"可恶的现实主义"这一建筑理念，《当今欧洲建筑中可恶的现实主义》，《设计书评》第17期（1989年冬），第17~20页。
2 勒默·凡·托恩，《反建筑之建筑学》，CTHEORY，网络出版，刊载于1997年9月24日，于2012年9月15日检索，http://www.ctheory.net/articles.aspx?id=94#note22。
3 同上。
4 同上。
5 考虑1994年Any公司在蒙特利尔举行的讨论会至关重要，该公司由彼得·埃森曼和辛西娅·戴维森管理的智囊团最终出版了《Anyplace》一书。
（辛西娅·戴维森，《Anyplace》，麻省理工学院出版社，1995年）。也可以在屈米的《介入》（第229页）以及詹姆逊的《空间具有政治性吗？》（弗雷德里克·詹姆逊，《空间具有政治性吗？》）（第196页）找到相关的讨论意见。
6 穆雷·弗雷泽，《文化语境下的批判性建筑》，《建筑学报》，第10卷，2005年，第317~322页。
7 辛西娅·戴维森，《Anyplace》，麻省理工学院出版社，1995年，在一次讨论中做的评论，Cf.C.C,第234页。

"constraints" are here interpreted as the complex spatial matrix generated by the existing urban fabric, which was "anchored in street, square and block", as the architects explain. The covered surfaces and patios originate in the previous configuration and, by the same token, bear the formal and spatial traces of that city area previously characterized by semi-industrial sheds.

Capelinhos Vulcano Interpretation Center, in Capelinhos, Azores, Portugal, is a clear materialization of the intention to trace the time of buildings and landscapes. The overall complex is intended to remember, explain and testify to the significant transformations that the area of the lighthouse of Capilhnos has undergone following the 1957–1958 eruption. The new project copes with the "existing" and its history via the premise of preserving the ruins and reconsidering the modified landscape. Ruins of the previous settlements are hence crucial for the development of the new Interpretation Center. More than limitations, the ruins are considered a fundamental occasion for making a "journey through history, subdividing the museological content of an informative, scientific and leisure character into three distinct sections: before, during and after the eruption", explain the architects.

The burdensome cultural and historical heritage carried by El Albaicín (one of the historical neighborhoods of the city of Granada, Spain) represents the main constraint that García Torrente Arquitectos encountered while re-designing the *Zayas House*. The project, in fact, is a renovation of a sixteenth century Morisco house, typical in its typology of the courtyard house in the Arab-Moorish tradition. The new project has been tailored around the existing structures, spaces and details in order to combine these with new requirements. More than being limited or constrained, the *Zayas House* is literally guided by the presences of its past. To this extent, what might have appeared as a restrict condition has eventually led to unexpected results and a significant improvement of the previously abandoned and uncared-for architecture. Silvio Carta

1. A succinct version of Lefaivre's article in which she proposed the idea of "Dirty Realism" in architecture appeared as Liane Lefaivre, "Dirty Realism in European Architecture Today", *Design Book Review* 17 (winter 1989), pp.17–20
2. Roemer van Toorn, *Architecture Against Architecture*, CTHEORY, online publication, published on 1997.9.24, retrieved in 2012.9.15, http://www.ctheory.net/articles.aspx?id=94#note22
3. Ibidem.
4. Ibidem.
5. It is crucial to consider the symposium held in Montreal in 1994 by the Any Corporation, a think-tank run by Peter Eisenman and Cynthia Davidson, which eventually produced the book Anyplace. (Cynthia Davidson, *Anyplace*, The MIT Press, 1995). Among others, comments relevant to this discussion can be found in Tschumi's *Intervention* (p.229), as well as in Jameson's *Is Space Political?* (Fredric Jameson, *Is Space Political?*, (p.196).
6. Murray Fraser, "The Cultural Context of Critical Architecture", *The Journal of Architecture*, Vol. 10, 2005, pp.317–322.
7. Cynthia Davidson, *Anyplace*, The MIT Press, 1995, A comment made during a discussion, Cf. C.C,p.234

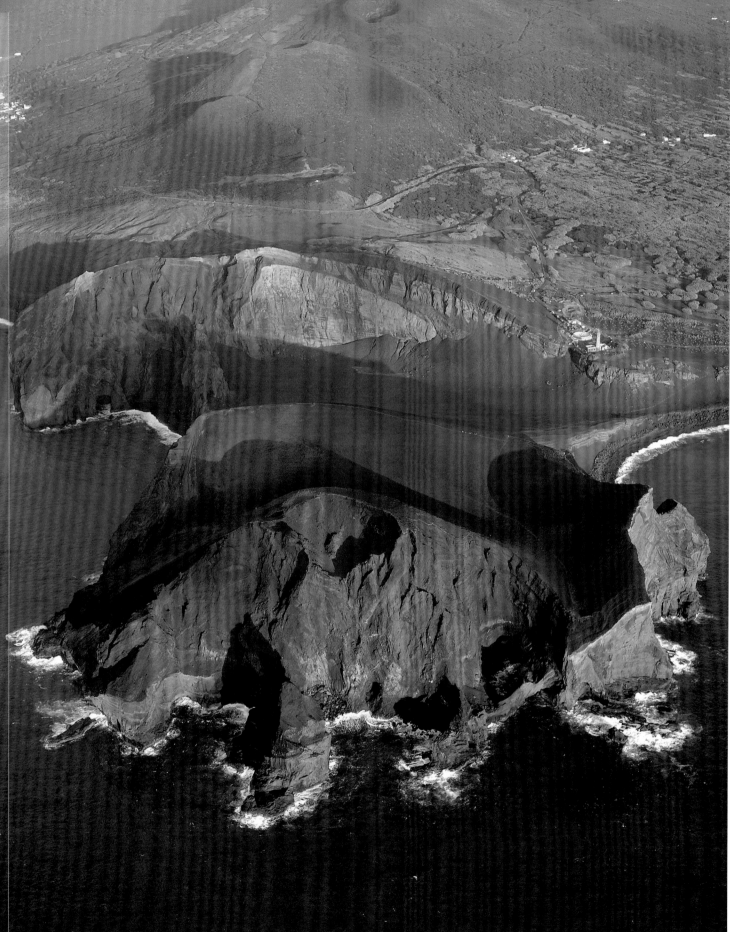

城市设计：从限制到优势 Urban How Constraints to Blessings

Capelinhos火山讲解中心
Nuno Ribeiro Lopes Arquitectos

　　1957至1958年法亚尔岛的火山大喷发掩埋了屹立于岛尖的灯塔，彻底改变了该岛的全貌，这一重要历史事件也解释了亚速尔群岛的成因，揭开了水下火山科学探索的重要篇章。

　　尽可能再现当时的壮观场面及保留喷发造成的后果应该成为即将在这片空地上建造的介入建筑的前提条件。建筑师意图建立一座讲解中心，因而灯塔及周围地区的改造应当保留废墟，同时恢复当时的地貌，使火山喷发后的状貌不可磨灭，并对灯塔至今为止的各个时期进行诠释。

　　因而，游客进入场地的道路就选在了较远的地方。游客远眺灯塔和火山时，也能够感受到它们，它们吸引游客走上间隔的石板路或者玄武岩鹅卵石路。路尽头最后几米的路程给游客的感官刺激会使其有一种瞬间穿越古今的感觉。

　　该中心位于地下，坐落于火山喷发前的地块，呈现了一趟穿越历史的旅行，该中心集知识性、科学性及休闲性于一体，分别对应火山喷发前、喷发中及喷发后三个时期的特征。

　　这一历史性旅行始于一个直径为25m的巨大圆形门厅，该门厅坐落于当时的一条巷道里，旅行的终点是灯塔的遗址，遗址罩在钢化玻璃箱里；灯塔有单独的入口，它的主要功能是用于瞭望。

　　该中心由钢筋混凝土建造而成，没有任何饰面，其功能主要是作为博物馆和道路旁边的建筑场景之间的互动元素。

　　遵循这些原则使Capelinhos火山讲解中心这座传统介入物产生的影响最小化，从而使游客可以从情感到科学等不同层面去领悟它。

　　人们可以参与展品的定义、概念、设计和选择使该建筑得到系统性调整，它将设计与建造活动完美地融合于一个连续的博物馆中，在这里新旧观念得以交融。不同的建馆专家选择了不同的展品，引入了他们的概念和作品所固有的技术和艺术手段而产生的变体。

　　对宏观变化和微观细节的处理体现了建筑师高水准的敬业精神，同时也使他们拥有大好的机会来实现建筑、展示内容和场地三者完美的统一性。

Capelinhos Volcano Interpretation Center

Important as a historic moment and as an aid to comprehending how the archipelago of Azores was formed, the eruption occurred in 1957-1958 definitely transformed the landscape of the island of Fayal, burying the lighthouse that stood on the tip of the island and opening up an crucial page in the scientific understanding of underwater volcanoes.

The furore of that moment and the consequences should be the premise in the intervention that would fill the existing lacuna. With the intention of setting up an interpretive centre, the transformation of the lighthouse and its surrounding area should preserve the ruin, recover the landscape of the area, make the resulting image unforgettable and provide an explanation of all the phases involved from the construction of the lighthouse until today.

Thus, the visitor's path to the site was chosen at a more distant venue; looking at the lighthouse and the volcano, which one sees and feels, the visitor is invited to walk, either on spaced flagstones or basalt cobblestones, the last few metres separating him from the beginning of a trip through time, simultaneously past and future.

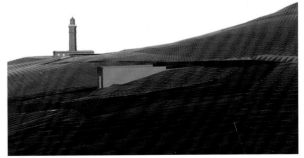

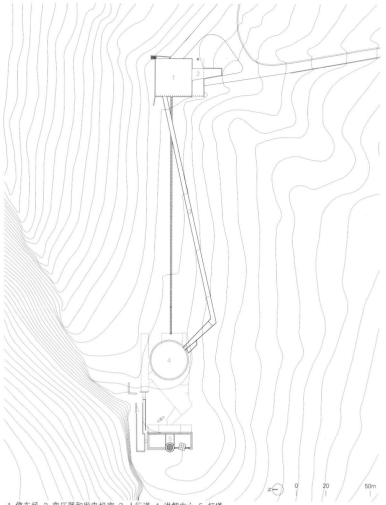

1 停车场 2 变压器和发电机室 3 人行道 4 讲解中心 5 灯塔
1. parking lot 2. transformer and generator house 3. pedestrian walkway 4. interpretation center 5. lighthouse

1903

1957—1958

2005

The building is subterranean, set on pre-eruption land, and provides a journey through history, subdividing the museological content of an informative, scientific and leisure character into three distinct sections: before, during and after the eruption.

The historical journey begins in the large, circular foyer, which measures 25 metres in diameter that stands exactly on the cul-de-sac that then existed and finishes on the ruins of the lighthouse, emerging in a steel-reinforced glass cube; the lighthouse has an independent entrance, its main objective is to function as a lookout.

Built in reinforced concrete and devoid of any facing, the building serves as an interactive element between the museum and the architectonic scenarios that exist along the walk.

The adoption of these principles minimizes the impact that a traditional intervention could imply, thus allowing the visitor to take in the place in stages, from the emotional to the scientific.

The possibility for people to participate in the definition, conception, design and choice of exhibits allowed the systematic adjustment of the building, transforming the act of projecting and constructing into a continuous laboratory where new values were conjugated with the usual ones. Different specialists chose the exhibits, introducing variants that resulted from technical and artistic means inherent in their conception and production.

This chance to control the most profound change or the slightest detail raised the level of the architect's responsibility and offered greater opportunity to fulfil the ideal of unity between the building, its contents and the site. Nuno Ribeiro Lopes Arquitectos

项目名称：Capelinhos Volcano Interpretation Center
地点：Capelinhos, Fayal Island, Azores, Portugal
建筑师：Nuno Ribeiro Lopes
项目团队：Sara Moncaixa Potes, Manuel Baião(Draftsman)
结构及维修工程师：Mário Veloso
设备工程师：Luís Andrade
电气工程师：Henrique Leal
质量监理员：Prospectiva, projectos, serviços, estudos, Lda.
甲方：Azores Regional Government Regional Secretariat for the Environment and the sea
建筑面积：2,000m²
造价：EUR 3,000,000
竣工日期：2008
摄影师：©Manuel Ribeiro (courtesy of the architect) - p.90~91, p.94, p.97, 98~99, p.100 left
©Filipe Jorge (courtesy of the architect) - p.95, p.100 right, p.101 bottom
©Sara Moncaixa Potes (courtesy of the architect) - p92(except as noted), p.93, p.101 top
©José Carlos Silva (courtesy of the architect) - p.88

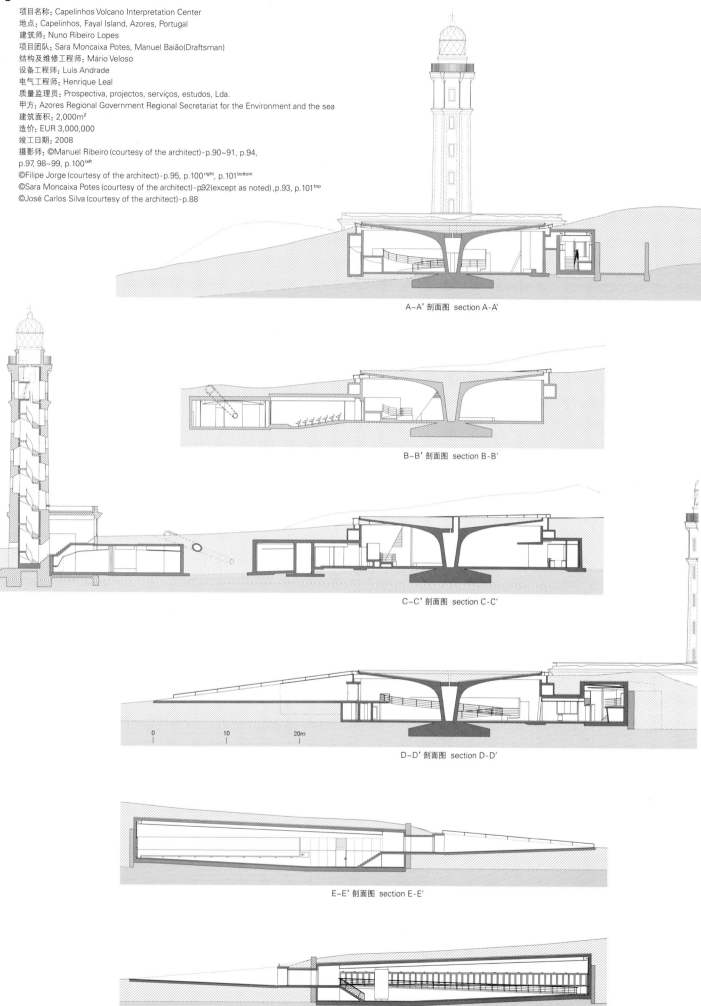

A-A' 剖面图 section A-A'

B-B' 剖面图 section B-B'

C-C' 剖面图 section C-C'

D-D' 剖面图 section D-D'

E-E' 剖面图 section E-E'

F-F' 剖面图 section F-F'

圆形屋顶详图 circular roof detail

1. PVC waterproof layer
2. bar 50x5
3. edge in bent plate 2mm thick
4. stainless steel flashing
5. laminated glass 10mm thick
6. stainless steel clamp
7. stainless steel tube
8. notifier ref. 1345 magnetic door holder
9. three coats of rubber paint with fiberglass mesh
10. "PUNTpart" glass clamp
11. eventual ash deposit
12. neoprene layer

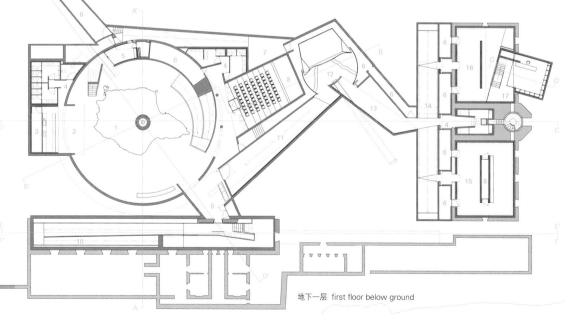

地下一层 first floor below ground

1 前厅	1. foyer
2 吧台	2. bar
3 仓库	3. store
4 卫生间	4. wash rooms
5 经理办公室	5. management office
6 技术区	6. technical area
7 控制中心	7. control room
8 礼堂	8. auditorium
9 售票处	9. ticket office
10 临时展厅	10. temporary exhibition
11 灯塔	11. light houses
12 火山喷发	12. eruption
13 火山	13. volcano
14 世界火山简介区	14. volcanoes of the world
15 亚速尔群岛展区	15. Azores
16 法亚尔岛展区	16. Fayal
17 商店	17. shop

1 入口
2 前厅
3 技术区
4 临时展厅
5 临时展厅出口
6 讲解展区出口
7 露台
8 灯塔
9 圆屋顶通道

1. entrance
2. foyer
3. technical area
4. temporary exhibition
5. temporary exhibition exit
6. interpretative exhibition exit
7. patio
8. lighthouse
9. access to dome

一层 first floor

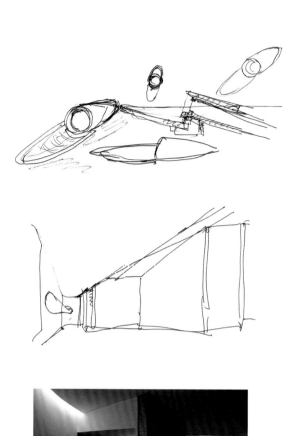

1. empty (e), enabling the alignment with the top of the volcano
2. glass clamps for posterior welding of the rims that are meant to fix the callotes that recoat the steel tube
3. glass clamps for posterior welding of the pre-rim which is meant to fix the rim that assembles the glasses

构筑墙体 building the walls

1. the interior side of the tube aligns tangent with existing wall
2. the slight difference between the diameters of the mold and the tube enables a final tuning to adjust the monocle
3. rims meant to fix callotes that recoat the steel tube enter simultaneously with the tube

管道的接合 cementing of the tube

1. movable grid
2. rim, sealing agent and tempered glass
3. hydrofuge plasterboard
4. callotes to recoat tube
5. final piece to recoat rim and glass
6. rim, sealing agent and tempered glass

饰面 finishing

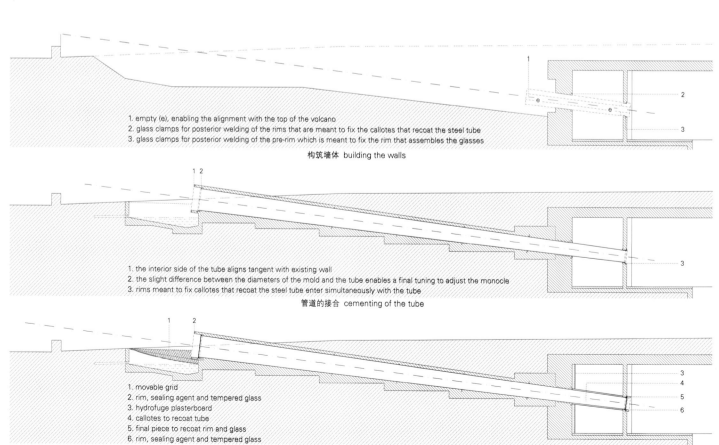

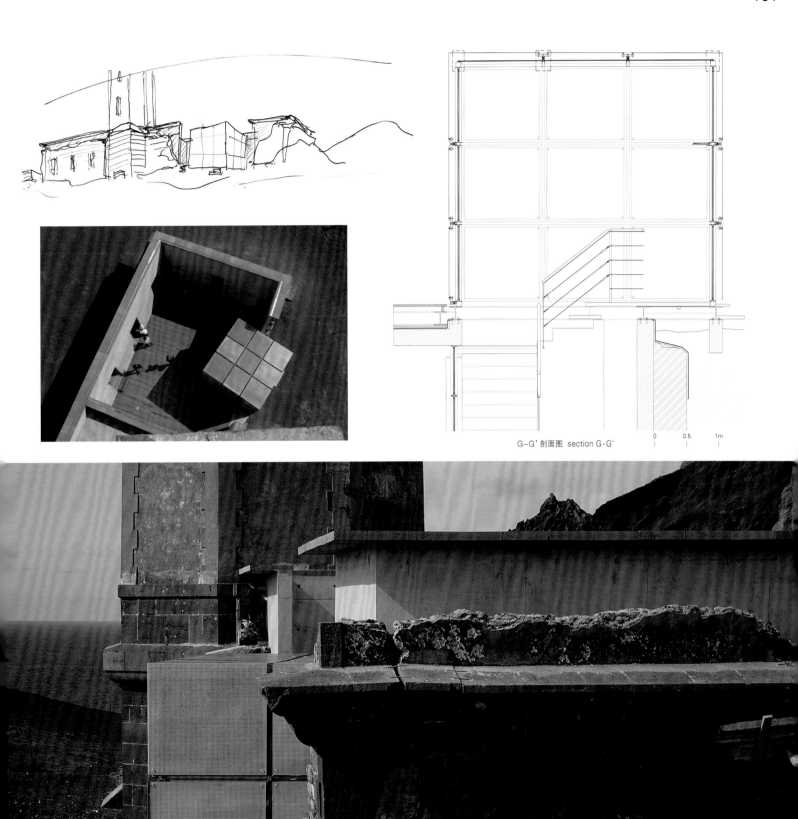

G-G'剖面图 section G-G'

Fontinha码头

Alexandre Burmester Arquitectos Associados

Fontinha码头的仓库沿杜罗河的河滨路分布，俯瞰这条河流。考虑到路宽，它们几乎形成了河流的一部分。这些仓库坐落于D. Luís桥和Arrábida桥之间的Fontinha 码头，与后面的一条窄路Rua Viterbo de Campos形成一个夹角，也是由Escadas das Fontaínhas的台阶进入。

总的来说，这些仓库残破不堪，只有墙面和结构进行过修缮。重建计划的主要目的是充分利用现有的布局及建筑物，通过修葺使其面貌焕然一新。

鉴于这些仓库与杜罗河及周围环境的关系，重建过程中每一块空间的开发与占用都必须与杜罗河紧密联系，哪怕仅仅是通过它所能看到的景色来联系。这些建筑物潜在的用途是通过混合来发展多功能性，进而吸引不同人群的眼球。建筑师们也因此创造了商业价值和居住空间。

重建这些仓库所用到的材料与原始材料一样，从而保留了仓库的部分原貌。

仓库的主体框架由钢筋混凝土建造，上层为金属结构，新的屋顶则是由胶合木材建造的。这些设施充分彰显了对品质与现代外观的双重追求。

这一自然的混合对工程质量要求非常高，不仅是对空间的开发，还要求加入现代化的元素。在施工的过程中已经考虑到了上述这些因素，也考虑到了与周围环境的融合。提案中的空间及其形式力求体现上述概念。

一方面，修葺好的建筑物自身浑然一体，另一方面，新增加的元素打破了传统的束缚，使人们的眼睛为之一亮。

Fontinha Wharf

The warehouses are on the riverside road looking out over the River Douro. Given the width of the road, they are almost part of the river. They are located at Cais da Fontinha ("Fontinha wharf"), half way between the D. Luís Bridge and Arrábida Bridge. They form an angle with Rua Viterbo de Campos, a narrow road running behind them, and are also served by the "Escadas das Fontaínhas" steps. In general, the buildings were highly dilapidated and only the walls and structures have been renovated. The main goal of the

东北立面 northeast elevation

西北立面 northwest elevation

西南立面 southwest elevation

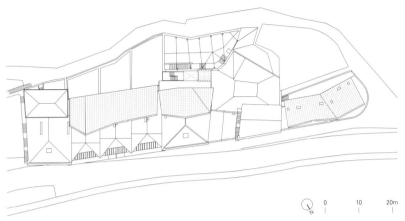

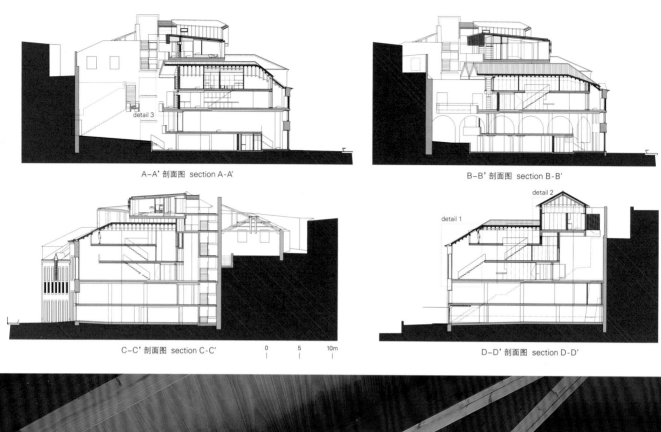

A-A' 剖面图 section A-A'

B-B' 剖面图 section B-B'

C-C' 剖面图 section C-C'   0 5 10m

D-D' 剖面图 section D-D'

plan was to take advantage of the buildings, by reinventing their forms and making the most of their location.

In their relationship with the river and their setting, it will be necessary to take into account that any space or use to be developed must always be associated with the river, even if only through the views that it offers. The buildings' potential uses were developed with a mixed in order to create multi-functionality so that they can attract diversified, complementary interests. Architects have therefore created spaces for business and housing.

The warehouses were rebuilt with the same type of old materials, following some of their original forms.

The main structure is made of reinforced concrete, while the structure of the upper floors is metal. The new roofs are made of laminated timber. The proposed finishing's show a concern for quality and a contemporary image.

A complex of this nature requires high-quality work. It is not enough to develop the space; it is also necessary to add a contemporary, modern image. This was taken into consideration during the development, along with respect for and integration into its setting. The proposed spaces and their forms seek to express these concepts.

On the one hand, the existing buildings have been restored to make them appear as a harmonized whole, and on the other hand additional forms have been introduced to break with tradition and make them stand out. Alexandre Burmester Arquitectos Associados

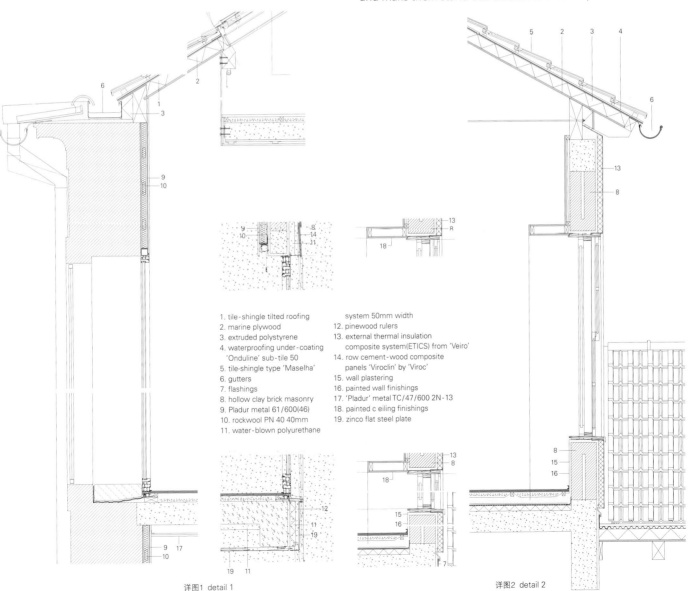

1. tile-shingle tilted roofing
2. marine plywood
3. extruded polystyrene
4. waterproofing under-coating 'Onduline' sub-tile 50
5. tile-shingle type 'Maselha'
6. gutters
7. flashings
8. hollow clay brick masonry
9. Pladur metal 61/600(46)
10. rockwool PN 40 40mm
11. water-blown polyurethane system 50mm width
12. pinewood rulers
13. external thermal insulation composite system(ETICS) from 'Veiro'
14. row cement-wood composite panels 'Viroclin' by 'Viroc'
15. wall plastering
16. painted wall finishings
17. 'Pladur' metal TC/47/600 2N-13
18. painted ceiling finishings
19. zinco flat steel plate

详图1 detail 1   详图2 detail 2

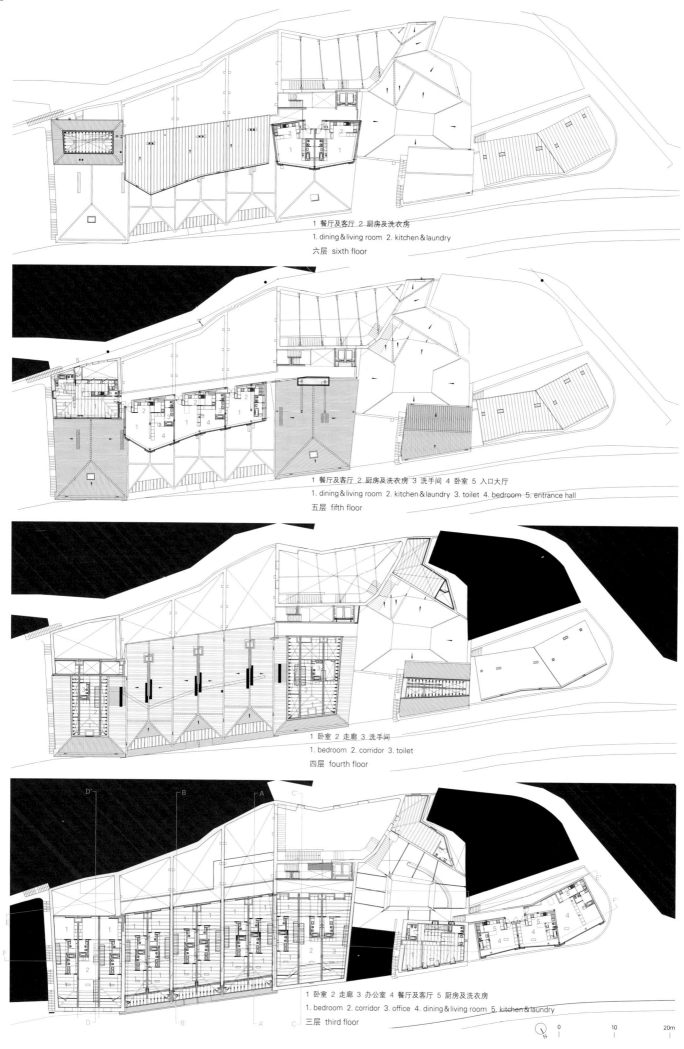

1 餐厅及客厅 2 厨房及洗衣房
1. dining & living room 2. kitchen & laundry
六层 sixth floor

1 餐厅及客厅 2 厨房及洗衣房 3 洗手间 4 卧室 5 入口大厅
1. dining & living room 2. kitchen & laundry 3. toilet 4. bedroom 5. entrance hall
五层 fifth floor

1 卧室 2 走廊 3 洗手间
1. bedroom 2. corridor 3. toilet
四层 fourth floor

1 卧室 2 走廊 3 办公室 4 餐厅及客厅 5 厨房及洗衣房
1. bedroom 2. corridor 3. office 4. dining & living room 5. kitchen & laundry
三层 third floor

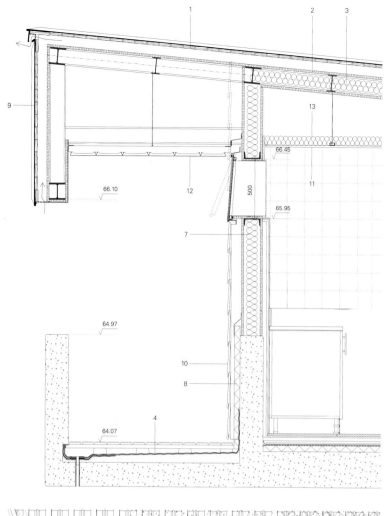
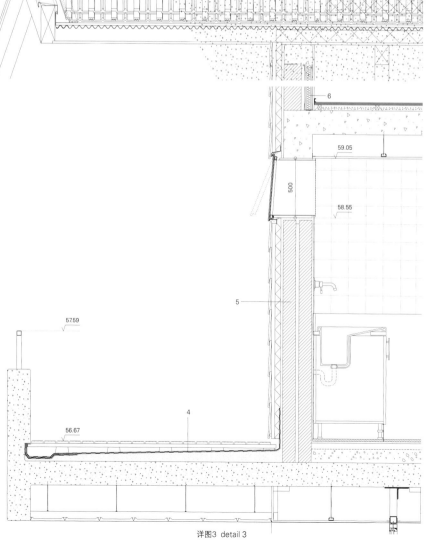

详图3 detail 3

1. zinc plated finish tilted roofing
2. raw cement-wood composite panels 'Viroc'
3. rockwool PN70 100mm
4. accessible flat roofing
5. hollow clay brick masonry
6. 'Pladur' metal 61/600 (46)
7. rockwool PN100 100mm
8. water-blown polyurethane system 50mm width
9. zinc plated finish wall cladding
10. raw cement-wood composite panels 'Viroclin' by 'Viroc'
11. 'Pladur' N-15-M-82/600 c/lã de rocha
12. ext. perforated natural anodized aluminum plates suspended ceiling
13. rockwool PN70 80mm

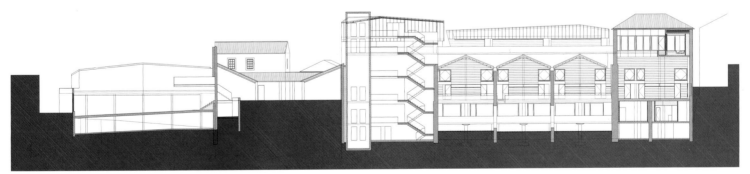

E-E' 剖面图 section E-E'

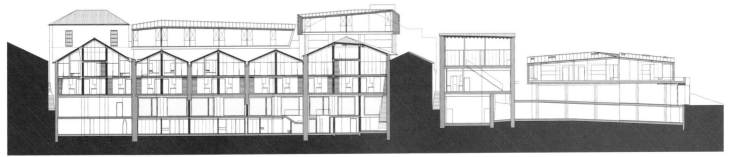

F-F' 剖面图 section F-F'

项目名称：Cais da Fontinha
地点：Margina de Gaia, Oporto, Portugal
建筑师：Alexandre Burmester, Andre Lopes Cardoso
合作方：Jose Carlos Goncalves, Fatima Burmester, Jorge Toscano
结构及水利设施：Rui Furtado, Rodrigo Castro
电气、机械及安全设施：Antonio Ferreira, Pedro Albuquerque
承包商：Luis Trocado Ribeiro
总建筑面积：6,000m² 造价：EUR 4,000,000
设计时间：2004 竣工时间：2010
摄影师：courtesy of the architect-p.102~103, p.115
©Marcos Oliveira(courtesy of the architect)-p.104, p.106, p.109, p.111, p.112, p.113, p.114~115

圣伊莎贝尔住宅

Bak Gordon Arquitectos

该项目最重要的一点或许就是要参考里斯本的城中城,城中城的建筑文脉已经深深地烙印在了街道、广场以及街区中。里斯本有很多这样的建筑,或多或少地复古,或高或矮地耸立,总显得难以琢磨。

通过修缮一些地区的脉络,这座被遗忘的破败之城可以得到修复,这种相互叠加的城市脉络能够促进城市建筑的复兴。

所有这些都与建于圣伊莎贝尔一个街区中部的那两座住宅项目有关,其用地面积为1000m²,以前用作半工业棚户区,穿过一个小商店就可以到街上。

该项目计划修建两座住宅,较大的一座用于家庭的日常生活,而另外一座只设有两间卧室,用于出租,这两座建筑均修建在400m²的审批用地中,代替了现有的棚户区。

这块地因建筑物周围突兀的空地而显得与众不同,同时由于其邻近建筑立面所形成的竖直的周边环境,因此需要一个相对水平的建筑

来与之形成对比。

所以，建筑师们建造了一座空间规则且极具层次感的住宅，中间是中空空间，周围设有可以举办各种活动的起居空间。前面的公共露台建于两座住宅之间。在这座住宅内，我们将漫步于露台、花园（一些花园可以用作冥想之所，而另外一些规模较大的花园可以得到有效利用）和即将生长在这里见证岁月流逝的树林中。

这座住宅几乎全是由裸露的钢筋混凝土建造而成，它与周围的分界均被绿色的攀爬植物所覆盖，这些植物改变了自然环境，而其余的墙体和屋顶则显得粗犷却又不失精致，来抵抗自然环境的影响。

游览完这些地方，封闭的建筑和开阔的空地会使人产生一种积极与消极相互对立的感觉，这种感觉主导着空间的规划方式。模块化的钢化窗户将屋内和屋外分隔开来，需要过滤清洗的地方相对较窄，而提供外界全景的部分则较宽。参观的入口是一扇黄色的门。

Santa Isabel Houses

Perhaps the most important thing in this project is the desire to refer to the city that exists within the city – the places inside the city, whose matrix anchored in street, square and block. There are many such places in Lisbon – more or less old, deeper or more open to the sky, but always very impenetrable.

This city, so often abandoned and unhealthy, can be recovered, giving way to another network of places, like overlapping meshes that can constitute a regeneration of the urban fabric.

All this concerns the project for two houses built in the midst of a block in Santa Isabel, a site with an area of about 1,000m² previously occupied by semi-industrial sheds and with access via a small store open to the street.

120

1 入口	1. entrance
2 厨房	2. kitchen
3 起居室	3. living room
4 办公室	4. office
5 卫生间	5. toilet
6 卧室	6. bedroom
7 浴室	7. bathroom
8 游泳池	8. swimming pool
9 花园	9. garden
10 停车场	10. parking
11 餐厅	11. dining room
12 大厅	12. hall
13 食品贮藏室	13. pantry

一层 first floor 屋顶 roof

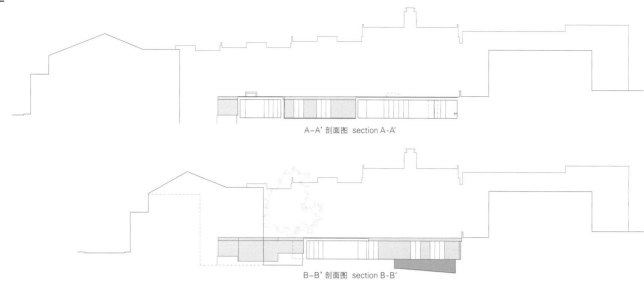

A-A' 剖面图 section A-A'

B-B' 剖面图 section B-B'

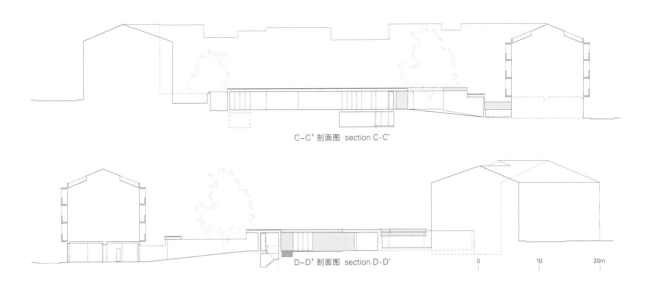

C-C' 剖面图 section C-C'

D-D' 剖面图 section D-D'

The program mandated the construction of two houses, a bigger one meant for the family's daily life and another two-bedroom one to be rented – all in the area of about 400m² for which construction was authorized, replacing the existing sheds.

The site was notable in that the empty space stood out with respect to the built, and for the vertical surroundings embodied in the facades of the neighboring buildings, which would suggest a very horizontal building, in contrast.

So the architects built a house with very regular and hierarchic spaces – the voids – around which the programmatic living spaces gravitated. A first patio, more public, receives and distributes between the two houses. Inside the house, we move among patios and gardens (some more contemplative, others bigger and for effective use) and trees which will grow here, projecting the scale over time.

The house is almost obsessively built solely of exposed reinforced concrete. Peripheral limits are covered in green climbers (changing natural element), while the other walls and roofs are left as such, simultaneously powerful and delicate, to resist the pressure of the environment.

Throughout these places an illusion is created in the confrontation of positive/negative, closed construction and void, which directs how the space is structured. Between "being inside" and "being outside" are the modular steel windows, less wide where filtration is desired and larger to provide a generous expanse. Whoever goes there must enter by a yellow door. Bak Gordon Arquitectos

项目名称：2 Casas em Santa Isabel
地点：Lisbon, Portugal
建筑师：Ricardo Bak Gordon
设计团队：Ana Durão, Nuno Costa
设备工程师：Gonçalves Pereira, Natural Works
景观设计：FC, Arquitectura Paisagista
总承包商：686-Construções Lda.
甲方：Tiago e Paula Viana
用地面积：1,261m²
建筑面积：560m²
施工时间：2007—2010
摄影师：©FG+SG Architectural Photography

安嫩代尔住宅
CO-AP

安嫩代尔住宅建于一块狭长地带（1：8.5），是市中心具有维多利亚风格的二层联排住宅，这次项目包括对其进行改建和扩建。甲方的要求是对原来的二层联排住宅进行再装修，进而设计成一座全新的四卧室住宅，拆除一个摇摇欲坠的带有波纹金属覆层的单坡屋顶，同时还要拆除一座曾经是圣诞装饰品加工厂的巨大金属棚屋。

该基地的西南面紧挨着住宅区，北面则与改建的钢琴厂仓库紧紧相连。从东西两个方向可以分别观赏到许多种植在此的树木以及绿色的街景。

将以上这些因素都考虑进去，联排住宅后部的新扩建部分应凸显出该基地的长度，新的楼层高度要符合自然地势，因此一系列的错层与小型庭院就应运而生，位于其北面的8m高的钢琴厂的阶梯式墙体正好成为其背景。

该地方浓荫遮蔽，使建筑师们有机会去探索如何对玻璃加以不同利用。开敞的窗户颠覆了传统习惯，采用的是实心嵌板，而墙体却使用了透明玻璃，使人们一下就能看到庭院。屋子南面的天窗起到了控制阳光的作用，它使屋子一年四季都可以充满日光，主庭院旁的全玻璃式画廊过道能够捕捉到冬日里的阳光，因而成了该住宅其他地方的热辐射源。

二层扩建部分中的主卧从南面进行了斜移，尽可能不遮挡相邻院落的阳光。此外，该房间背向邻近的住宅单元，因而有更多的私人空间，减少了被人窥视的问题。

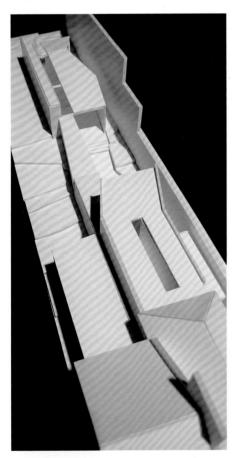

该项目对材料和饰面的选择都考虑了后面钢琴厂砖墙这一因素，将现有的刷过漆的砖石刮掉，裸露出里面的砖及压缩水泥板覆层。新的大方石抹灰层不用上漆，而金属屋顶板是镀锌的，可以任岁月侵蚀。钢材漆上血红色，使人想起原钢底漆的颜色。拆下来的砖可以再利用去铺设人字形花纹的行车道。

该项目的主要创意就是要使本住宅和附近钢琴厂的墙有紧密联系。这座房子通过使钢琴厂的墙也融入该基地中，从而与该墙的材料性与庞大性形成呼应。建筑师们负责任地对现有联排住宅进行了调整。他们没有对现有联排住宅的内部进行大的改造，这样有利于将睡眠区域设置在现有房间内，建筑师们仅恢复了联排住宅原有的外部特征，从而使其与周围街景相协调。

这一项目让人们有机会探索另一种建筑风格，使市中心的这座两层联排住宅更现代化。这块基地的限制条件及其所提供的机会、项目大纲以及市政规划条款共同造就了一座独一无二的房屋，它与其所在的基地形成了一种深层联系，并最终成为一个具有实用性兼舒适性的住宅。

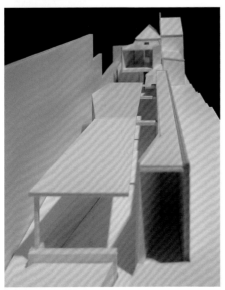

Annandale House

The project consists of alterations and additions to an existing inner-city two-storey Victorian terrace house built on an exceptionally long and narrow site (ratio 1:8.5). The client's brief was to design a new four-bedroom residence by refurbishing the original two-storey terrace and demolishing a shamble of corrugated metal clad lean-tos, and an enormous metal shed which was formerly used as Christmas decorations factory.

The site is closely bounded by residential units to the south-west and the Piano Factory warehouse apartment conversion to the north. The west and east offer views to established trees and green streetscape respectively.

Taking these issues into consideration, the new additions attach to the rear of the terrace house and unfold along the length of the site, with new floor levels corresponding to the natural topography, resulting in a series of split levels and pocket courtyards with the eight-meter high stepped piano factory heritage wall along the northern boundary as their backdrop.

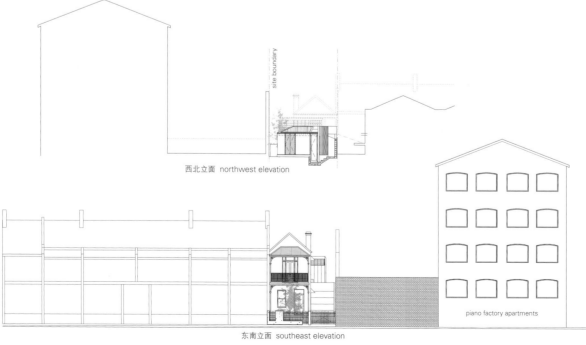

西北立面 northwest elevation

东南立面 southeast elevation

piano factory apartments

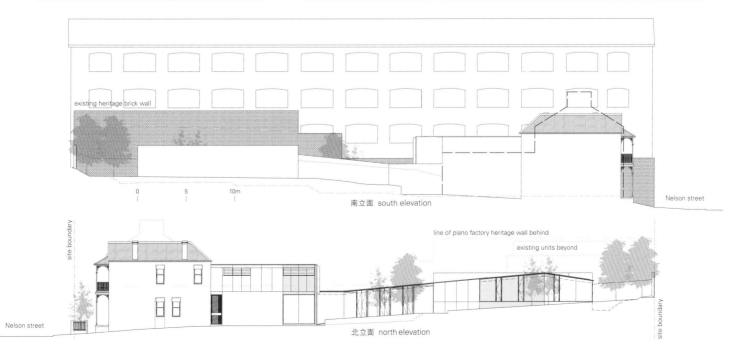

南立面 south elevation

Nelson street

line of piano factory heritage wall behind

existing units beyond

Nelson street

北立面 north elevation

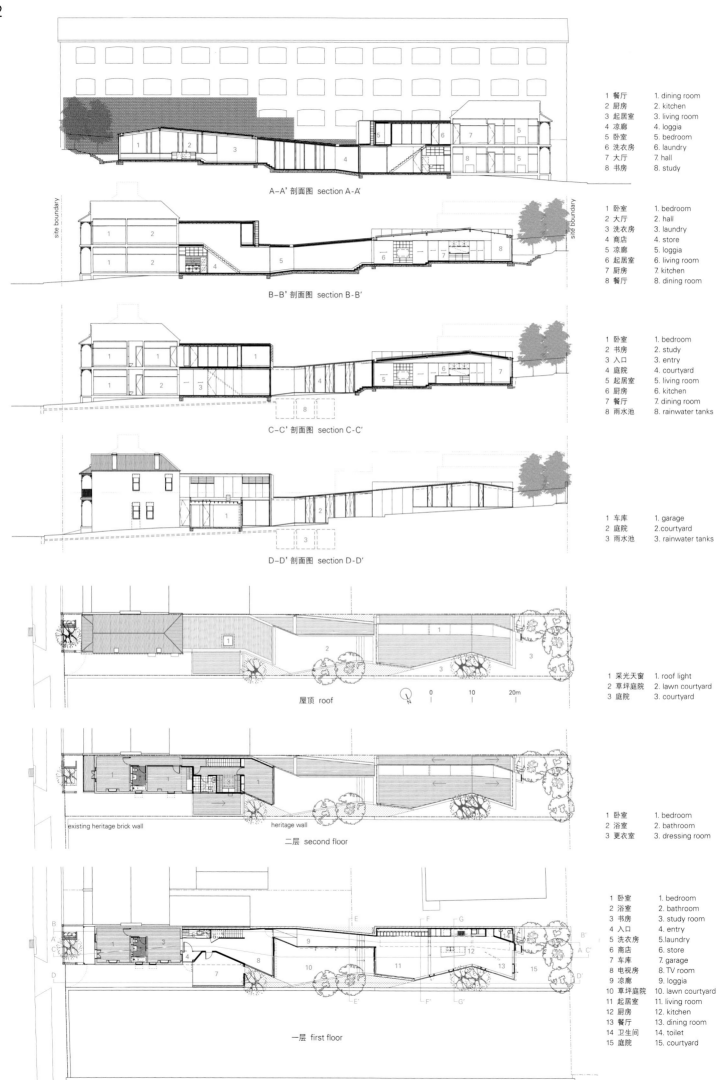

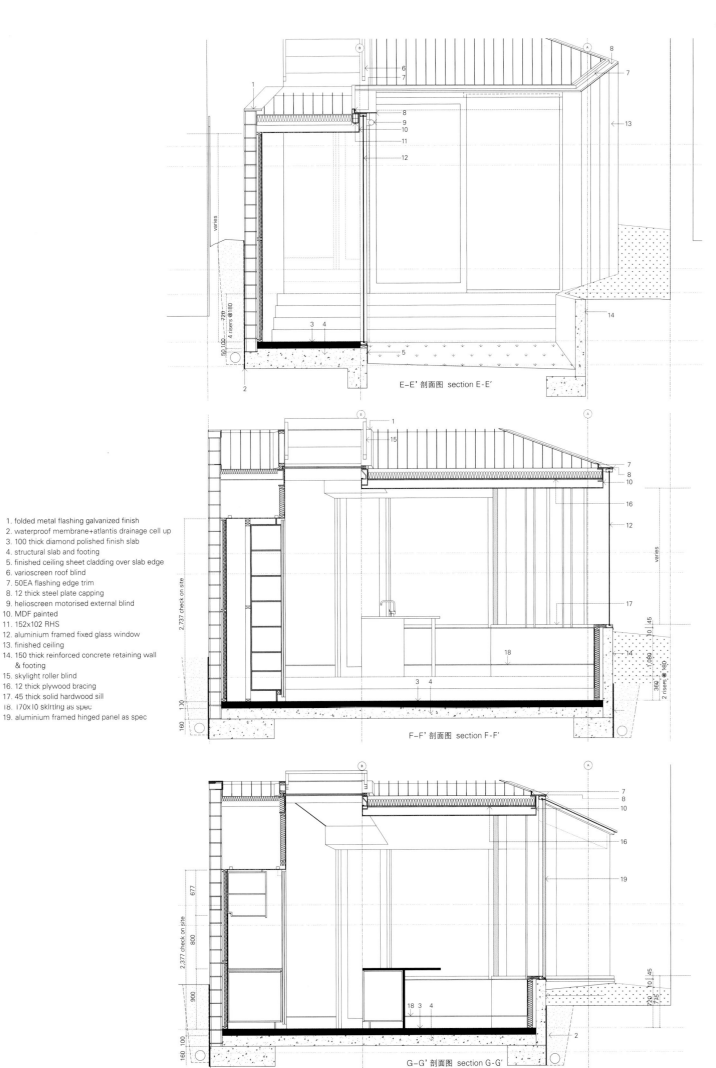

1. folded metal flashing galvanized finish
2. waterproof membrane+atlantis drainage cell up
3. 100 thick diamond polished finish slab
4. structural slab and footing
5. finished ceiling sheet cladding over slab edge
6. varioscreen roof blind
7. 50EA flashing edge trim
8. 12 thick steel plate capping
9. helioscreen motorised external blind
10. MDF painted
11. 152x102 RHS
12. aluminium framed fixed glass window
13. finished ceiling
14. 150 thick reinforced concrete retaining wall & footing
15. skylight roller blind
16. 12 thick plywood bracing
17. 45 thick solid hardwood sill
18. 170x10 skirting as spec
19. aluminium framed hinged panel as spec

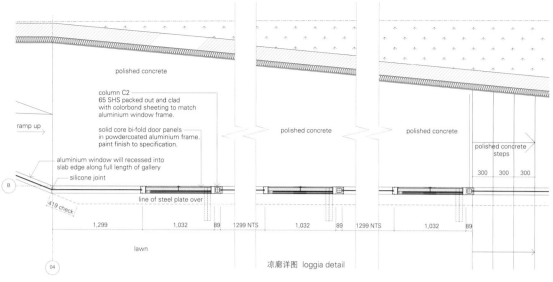

凉廊详图 loggia detail

The heavily shaded site gave the opportunity to explore the use of glazed elements differently. Inverting normal conventions, opening windows are solid panels and walls are clear glass, providing a visual connection to the courtyards. Sun-controlled skylights along the southern edge of the house capture daylight throughout the year. The glazed gallery hallway beside the main courtyard captures the winter sun and becomes a radiant heat sink for the rest of the house.

The main bedroom within the second storey addition angles away from the southern boundary to minimize overshadowing onto adjacent properties. Furthermore the room is orientated away from the neighboring residential units, increasing privacy and reducing overlooking issues.

The selection of materials and finishes for the project were influenced by the presence of the Piano Factory brick wall. Existing painted masonry was stripped back to expose sandstock brickwork and compressed fiber cement sheet cladding was left raw. New ashlar render is also left unpainted, metal roof sheeting is galvanized and left to weather with age. Steelwork is painted a blood red color, reminiscent of raw steelwork primer. Bricks from the demolition works were re-used for the new herringbone paved driveway.

Developing strong relationship to the neighboring piano factory wall was a major design opportunity for this project. The house responds to the materiality and immensity of the wall by respectfully integrating it into the site. There is responsible adaptive re-use of the existing terrace house. Architects chose not to make major internal alterations to the existing terrace which was useful for programming the sleeping areas into the existing rooms. Original exterior features of the terrace house was restored in keeping with the character of the immediate streetscape.

The project has provided an opportunity to explore an alternative model for updating the inner-city two-storey terrace house. The constraints and opportunities provided by the site, the brief and council planning controls have generated a unique house which has a deep connection to its site and is ultimately a very functional and comfortable home. CO-AP

项目名称：Annandale House
地点：Annandale, Sydney, Australia
建筑师：CO-AP
结构工程师：James Taylor & Associates
水力学顾问：Whipps Wood Consulting
勘测员：Kevin Brown & Associates
环境工程师：Noel Arnold & Associates
规划：Lockery Planning & Development Solutions
总楼面面积：296.85m²
竣工时间：2011.12

Ceschi住宅

Traverso-Vighy Architetti

Ceschi住宅坐落在维琴察老镇，房屋与老镇得天独厚的地理位置相得益彰，同时实现形式与结构上的平衡。与古城墙相傍，如设计不独特则不能完美展现具有悠久历史的外体及与其都市文脉的紧密联系，同时，其地理位置的优越性也决定了室内设计注重功能的重要性。

挖掘地下室使人们注意到了这座古镇的其他城墙，这些城墙可追溯到12世纪，从而使这个项目变得更加复杂和引人入胜。

每每面临新的建筑挑战，建筑师们总是尽可能轻地穿行其间，尽可能少地在环境中制造不和谐因素，不管它是环境问题抑或是与本例类似的建筑问题。贯彻上述原则最终形成的项目具有以下特点：每处细节都得到仔细安排，很多部分都由提前安装好的部件组成，还有能够维持较好的结构独立性的建筑构件，同时，这些构件还能与周边环境进行互动。

一个新结构被嵌入到Ceschi住宅带檐槽的墙体上，通高的框架是由层压落叶松木搭建而成的，界定出了住宅的格局，同时与房屋自身的盒状结构相得益彰，能充分发挥极强的抗震作用。

所有部件均为预制的，使工期和安装过程合理化。与此同时，这还要求设计和施工都具有极高的精确度。

地下室和一层作为专业书房。照明设备在这一区域需要格外留意，其设计出发点是降低能耗，利用自然光，无需安装窗户，光线也可以照射到地面上，同时也烘托出外墙表面的不规则性。当然，外墙装饰材料的选择也花费了建筑师的大量精力，这些材料会使该住宅成为一座绿色建筑。

未雕琢的陶瓷板覆以石灰，与一种特殊的多层保温材料相结合，可以对住宅内部的卧室区起到极佳的防潮作用，而墙体表面和最终的饰面主要采用了松木板和大麻砖，它们的保温和隔音性能极佳。原始框架和具有历史意义的部分利用传统技术复原，包括石灰刷就的小餐厅，而起保护作用的产品主要用蜂蜡制成。整座建筑的目的是为了把能耗降至最低，达到一级能耗标准。

Ceschi House

Ceschi House is located in the old town of Vicenza, and it is precisely with the peculiarity of the place that it seeks to establish a bond, as well as a both formal and structural equilibrium. Its position by the ancient town walls is called for a particularly attentive design, where the need to preserve the historical outer shell with its close ties to the urban tissue has been related to the necessary of readapting the interior in functional key.

The excavations in the basement have brought to light other parts of the town walls, dating from the XII Century, making the project more complex and fascinating.

Every time the architects face a new architectural challenge they try to thread as lightly as possible, to create as little disturbance as possible in the environment they are working with, whether it is an environmental or, like in this case, an architectural one. The result of this philosophy is projects where every detail has been carefully planned, and where many parts consist of pre-assembled elements, architectural objects capable of maintaining a marked structural independence, and that at the same time interact with the surroundings.

A new structure has been inserted within the gutted walls of Ceschi house; a skeleton in laminated larch wood which reaches the full height of the building, defining its distribution and establishing a collaborative relationship with the container characterized by an outstanding anti-seismic performance.

All parts have been prefabricated, something which have been made it possible to rationalize all phases of construction and as-

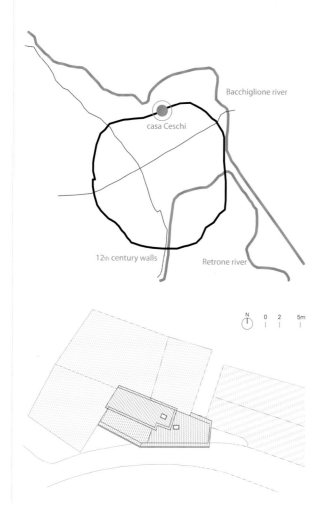

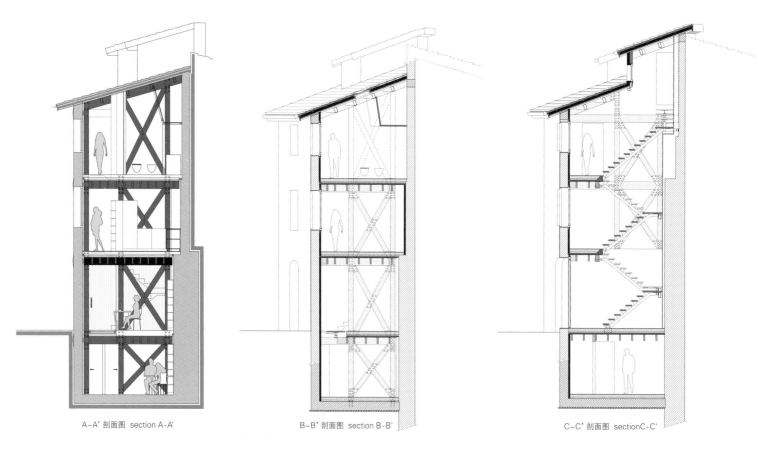

A-A' 剖面图 section A-A'　　　　　　B-B' 剖面图 section B-B'　　　　　　C-C' 剖面图 sectionC-C'

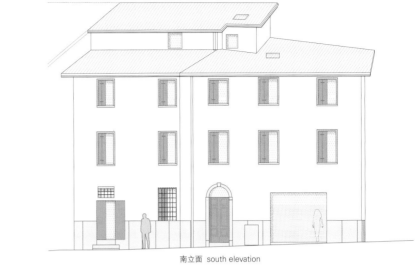

南立面 south elevation

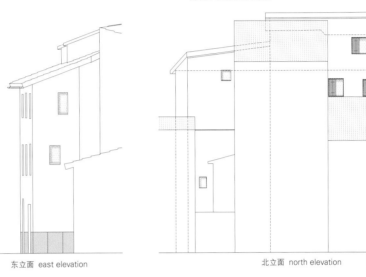

东立面 east elevation　　　北立面 north elevation

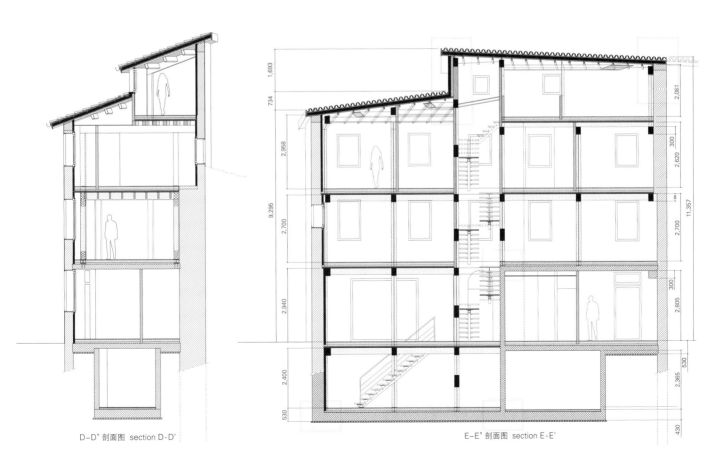

D-D' 剖面图 section D-D'　　　E-E' 剖面图 section E-E'

sembly; at the same time it has required a high level of precision both in design and execution.

The first floor and the basement will be used as professional study. The lighting design, ideated in order to reduce consumption and exploit the natural light, has been particularly attentive in this area, both to make the daylight reach the floor without windows and to enhance the irregular surfaces of the outer walls. Considerable attention has also been dedicated to the choice of the materials used to decorate the surfaces, and that contribute to make this a green building.

A special multilayer insulation has been combined with panels in untreated earthenware finished with lime, that would guarantee an excellent hygrometric control in the part of the interior used as dwelling, while the surface treatments and final touches have been done mainly with larch wood and hemp bricks, with an excellent performance in terms of thermal and acoustic insulation. The original shell and all the parts of historical value have been restored using traditional techniques, with lime-based diners and protecting products based on beeswax. The whole building has been designed so as to reduce consumption energy to a minimum, and it has been assigned energy consumption class A. Traverso-Vighy Architetti

项目名称：Casa Ceschi
地点：Vicenza, Veneto, Italy
建筑师：Giovanni Traverso, Paola Vighy
合作商：Giulio Dalla Gassa, Sheerja Iyer, Elena Panza, Valentina Rossetto
结构工程师：Ing. Franco Grazioli, Life engineering s.r.l.
施工方：Bios Edilizia
甲方：Barbara Ceschi a Santa Croce
用途：Private House
总楼面面积：290m²
造价：EUR 300,000
施工时间：2010.1—2010.5
摄影师：©Alessandra Chemollo (courtesy of the architect)

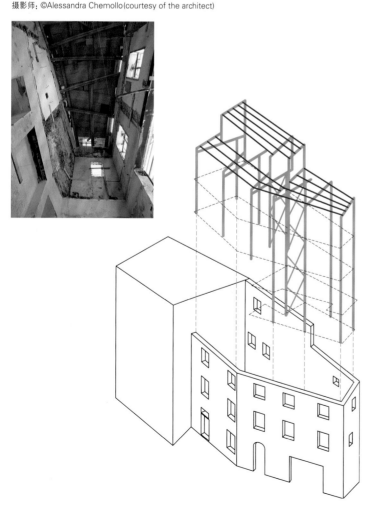

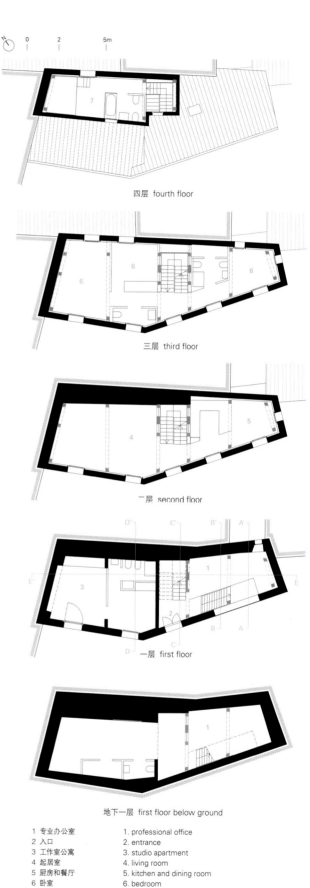

四层 fourth floor

三层 third floor

二层 second floor

一层 first floor

地下一层 first floor below ground

1 专业办公室 1. professional office
2 入口   2. entrance
3 工作室公寓 3. studio apartment
4 起居室   4. living room
5 厨房和餐厅 5. kitchen and dining room
6 卧室   6. bedroom
7 阁楼   7. attic room

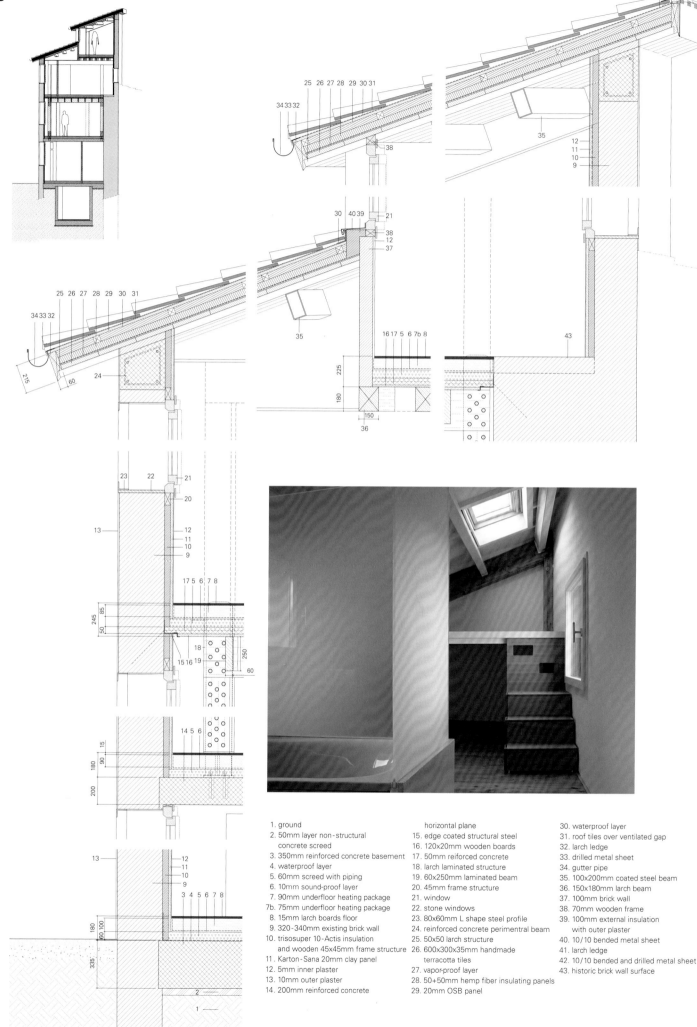

1. ground
2. 50mm layer non-structural concrete screed
3. 350mm reinforced concrete basement
4. waterproof layer
5. 60mm screed with piping
6. 10mm sound-proof layer
7. 90mm underfloor heating package
7b. 75mm underfloor heating package
8. 15mm larch boards floor
9. 320-340mm existing brick wall
10. trisosuper 10-Actis insulation and wooden 45x45mm frame structure
11. Karton-Sana 20mm clay panel
12. 5mm inner plaster
13. 10mm outer plaster
14. 200mm reinforced concrete horizontal plane
15. edge coated structural steel
16. 120x20mm wooden boards
17. 50mm reiforced concrete
18. larch laminated structure
19. 60x250mm laminated beam
20. 45mm frame structure
21. window
22. stone windows
23. 80x60mm L shape steel profile
24. reinforced concrete perimentral beam
25. 50x50 larch structure
26. 600x300x35mm handmade terracotta tiles
27. vapor-proof layer
28. 50+50mm hemp fiber insulating panels
29. 20mm OSB panel
30. waterproof layer
31. roof tiles over ventilated gap
32. larch ledge
33. drilled metal sheet
34. gutter pipe
35. 100x200mm coated steel beam
36. 150x180mm larch beam
37. 100mm brick wall
38. 70mm wooden frame
39. 100mm external insulation with outer plaster
40. 10/10 bended metal sheet
41. larch ledge
42. 10/10 bended and drilled metal sheet
43. historic brick wall surface

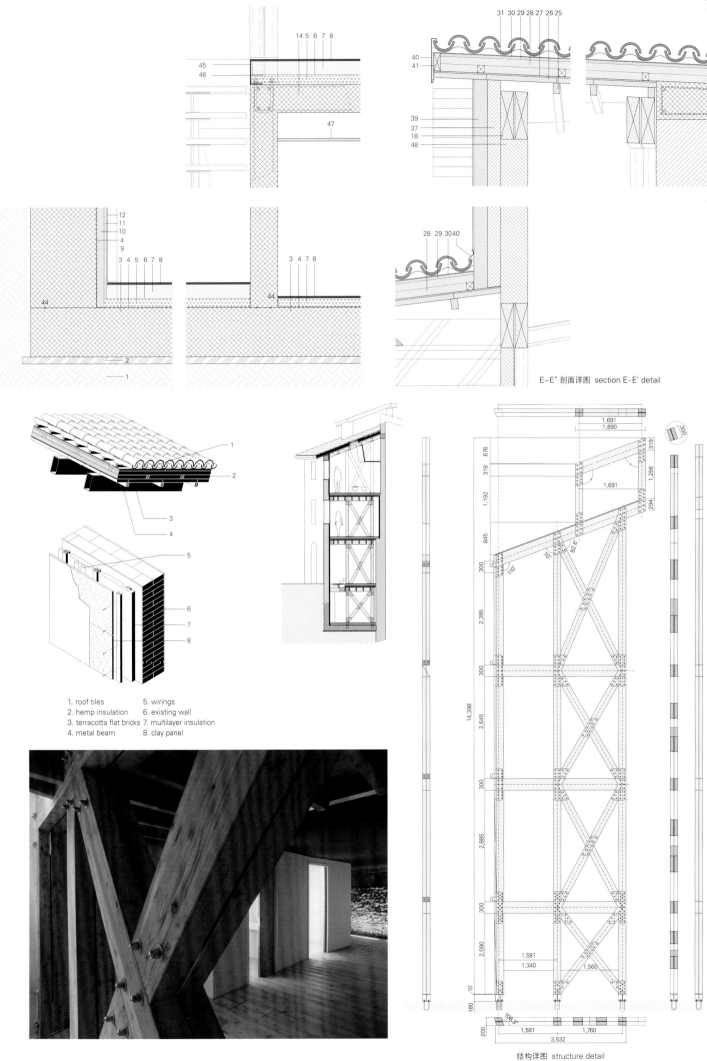

E-E' 剖面详图 section E-E' detail

1. roof tiles
2. hemp insulation
3. terracotta flat bricks
4. metal beam
5. wirings
6. existing wall
7. multilayer insulation
8. clay panel

结构详图 structure detail

Rizza住宅

Studio Inches Architettura

位于历史核心区的住宅

Rizza住宅位于瑞士提契诺州一个名为瓦卡洛的村落。该村有3000人口，过去靠农业为生，后来在20世纪逐渐发展成为一个居住区。该建筑位于城镇的"历史核心"区，其中部分住房可追溯到17世纪。Rizza住宅最早是作为一个旧谷仓，后来有一大家族定居此处，现在该房屋的一层成了该镇主要党派的会议场所。

尽管该项目意在充分利用6m×6m的室内空间，突出现有建筑的厚重砖墙所形成的巨型结构，但当地法律明令禁止擅自变更建筑的现有开口及其屋顶结构。

Rizza住宅呈方形、塔楼状，拥有6m×6m的空间，共四层。根据当地建筑法的规定，建筑的外围护结构维持原样：现有开口不做改动，屋檐高度不变，屋脊和坡屋顶的类型均不变。承重外墙是唯一一个没有变动的原有构件。室内彻底腾空重建，部分原因是因为墙体和地板存在安全隐患。

外观上，该项目欲凸显此建筑的历史背景，而室内则意在营造开阔的空间感，每层给人远远大于30m²的感觉。窗户和50cm厚的墙体的内表面齐平，以增加老墙的历史感。

客厅的双层高度增强了立体效果，也拓宽了空间。

轻质的金属楼梯四处可见，在视觉上交相呼应。

设备室和卫生间位于入口层，二层设有厨房和起居空间，三层有一个小房间、卫生间以及书房，在书房中可俯看到起居室，带有衣帽间和卫生间的主卧设置在顶层，从顶层可以将周围的门德里西奥峡谷尽收眼底，这片风景被愈演愈烈的城市扩张和土地荒废破坏了。建筑师相信对现有建筑进行改造是解决上述问题不得已而为之的好办法。

Rizza House

House in the Historical Core

Rizza House is located in Vacallo, a village of 3,000 inhabitants in the canton of Ticino, Switzerland. The town was originally a farming village, which gradually developed into a residential area during the twentieth century. The house is in the "historic core" of the town, with a portion of settlements dating back to the seventeenth century. Rizza House was an old farm storage, then residence for a large family and recently, only on the first floor, the meeting place for the main political party in town.
The project deals with the strict swiss laws that don't allow modifying existing openings and roof structure of the building. The project though wants to take out the most out of the 6x6m of the

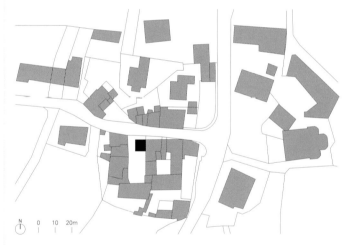

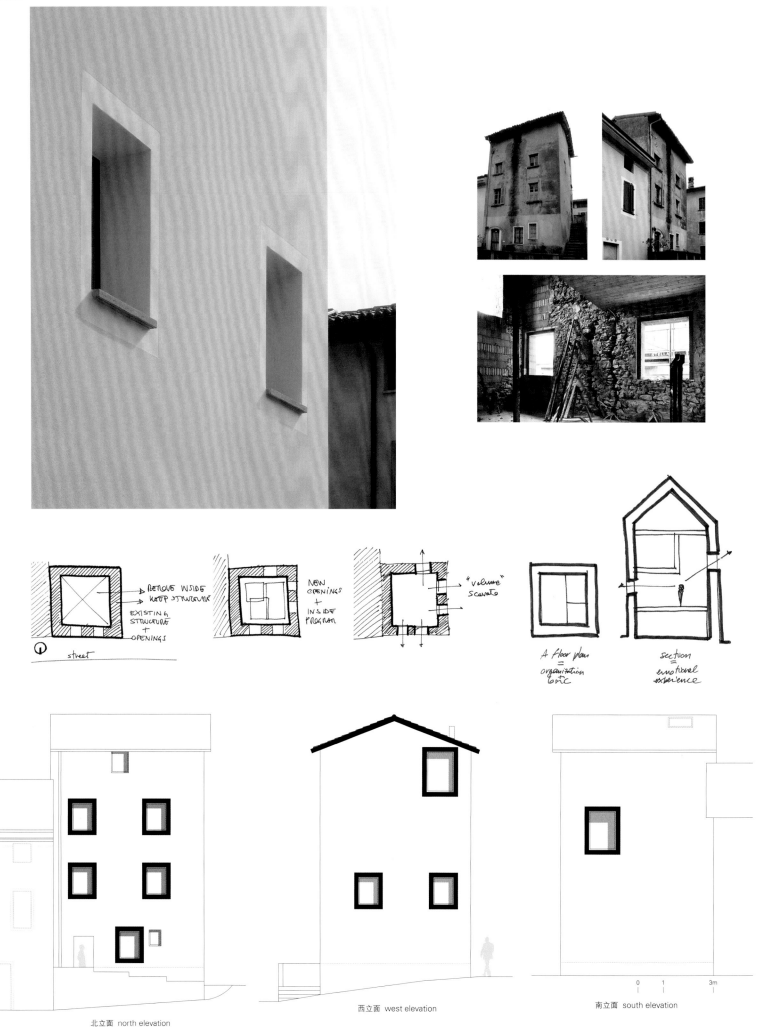

北立面 north elevation 西立面 west elevation 南立面 south elevation

interior spaces and to underline the massive structure of the thick walls of the existing building.

Rizza House appears as a square floor "tower-house" of 6x6m, on 4 levels. Due to restrictions set by local building laws, the envelope has not changed: the openings remained the existing ones, as well as the height of the eaves, the ridge and the type of pitched roof. The load-bearing exterior walls are the only pre-existing elements that remained untouched. The interior was totally emptied, partly because of the precarious condition of the walls and floors, and completely rebuilt.

The project wants to underline the historical background of the building on the outside while on the inside the intention is to give people the perception of a bigger space than the bare 30m² on each floor: the windows are flushed with the inner surface of the 50cm wall structure to increase the perception of the old walls.

The double height space in the living room increase the perception of verticality and size of the spaces.

The lightweight metal stairs are visible from all spaces which yearn for a visual relationship with each other.

Equipment room and WC are located on the entrance floor, kitchen and the living area are on the 2nd floor, the 3rd floor hosts a small room, WC and the study space that overlooks the living room, while the master bedroom is located on the top floor (with the cloakroom and WC) that overlooks the surrounding valley of Mendrisio, a landscape which is suffering from the increasing urban sprawl and waste of land. The architects believe that renovations of existing building could be a great and forced answer to that. Studio Inches Architettura

四层 fourth floor

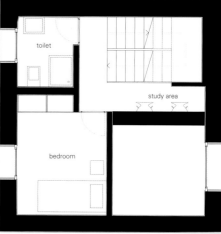

三层 third floor

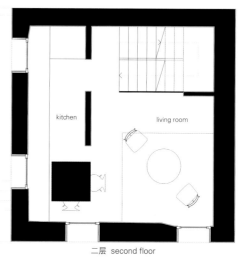

二层 second floor

一层 first floor

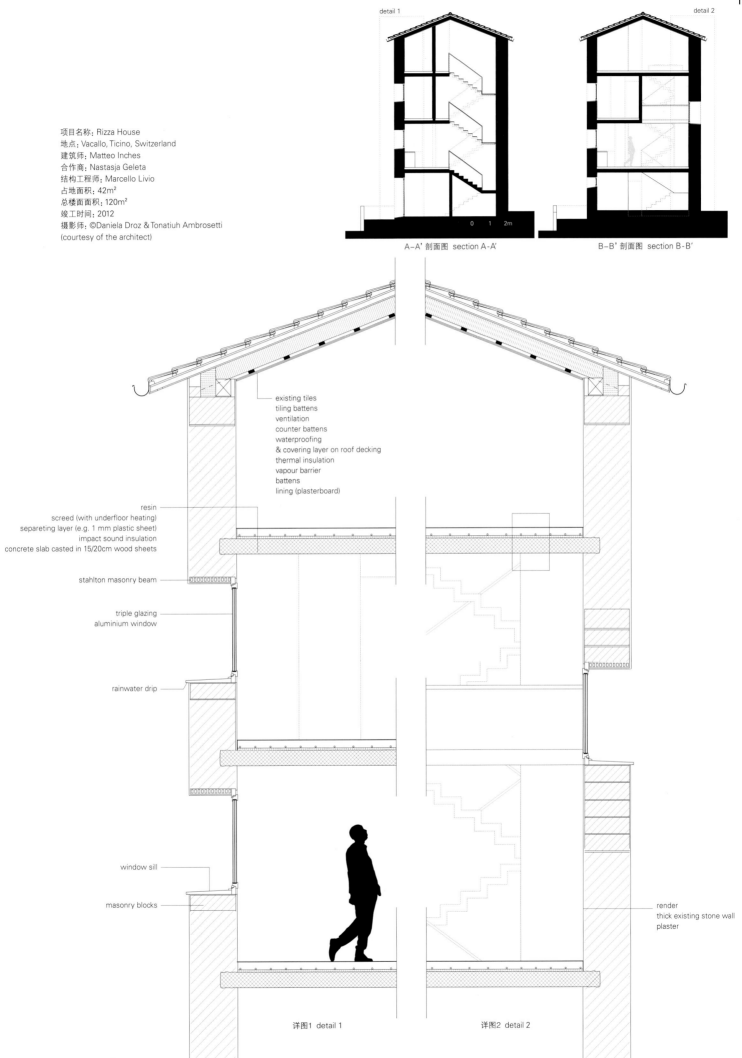

Zayas住宅

García Torrente Arquitectos

虽然早已破旧不堪，人去楼空，但是这座位于格拉纳达阿尔拜辛区的16世纪传统阿拉伯庭院住宅如今已成为一个家庭的住宅，该房屋曾经是皇宫，后来每况愈下，被非法入住。该建筑坐落在一个小型街区中。正面看房子有三层，对面两条街形成的斜坡使这座房子看起来处于一个复杂的楼层系统中，也正是环境使然，使得该住宅中的每一层都相互连通起来。

房屋周围布局包括院子，中心呈正方形，另外还有两条游廊相对排布。

楼梯、立面装饰、游廊，尤其是瞭望塔的外观使这座房子极具典型性。

该建筑最复杂也最具魅力的地方当属游廊。精致的建筑设计关系自然流畅，与庭院相辅相成，使人既感矛盾又倍觉丰富多彩。矛盾是因为房屋的内部和外部、公共和私人空间遵循双重理念；丰富多彩是因为它形成一个循环空间可以避雨，该空间可以作为集会和邻里聚会的场所、感光区、空气制冷或加热区或者小花园等。

一个非常敏感的区域可以完美捕捉到居住者的诸多情况。这个游廊界定了庭院和空地。

建在这块土地上的庭院要力求反映群居情况。房屋适宜的通风效果多亏了相互间的通透性。基本上来说，游廊和庭院是这个街区的灵魂，生动地刻画出了日常生活的点点滴滴。

Zayas住宅曾是阿尔拜辛区的贵族居所，由于时间和各种原因，它以前只分段出租而不像其他房屋那样整套出租。

修葺后，现在房屋已经是一个整体，但是里面住着八户人家，前任房主决定全部低价租给老人；住这么多人，多亏了房子的灵活性，可根据不同用途、家当多少、社区和不同家装风格等做适当改动。

建筑的潜力是指可根据不同时代的价值标准，对空间结构、功能、光线和形式等进行重新界定和布局调整。

该建筑仅由一系列的水平面和垂直面组成，墙体和地面搭建出了居住空间：庭院是个上空空间，没有内部结构，游廊成了其模糊的界

线。

　　建筑融抽象、有机与矛盾于一体。房屋的私人外围护结构呈现出简单的几何形，功能俱全，房子内部未分隔。一般来说，一个单独的区域或一系列两到三个区域都是由上空空间连接起来的，通常不使用木制品。这些空间可自由安排使用，不做限定，完全取决于各个家庭或一大家族的特定要求，这些都会随着时间变迁。具体用途可以通过添置家具来实现。

　　游廊和庭院可公用，方便邻里之间沟通。房内房外的关系更自由、灵活且多姿多彩，是幸福居所不可或缺的因素。

Zayas House

Although in a state of obvious deterioration and abandonment, this traditional Arab courtyard house from XVI century in the Albaicín district of Granada, originally conceived as palace, became home for family, reaching our days, after a continuous process of deterioration, illegally occupied. The building is situated in a small block. It has a front of three stories to two opposite streets whose slope has allowed the existence within a complicated level system that articulates the different floors of homes.

The distributor element around the house is the yard practically square centered and positioned with two galleries flown on opposite sides.

The stairs, the decoration of facades and galleries and especially the appearance of a lookout tower element, complete the typological sample of housing.

One of the most complex and rich elements of the building is the gallery. Its subtle constructive relationship so natural and posed with the court, makes an ambiguous and versatile space. Ambigu-

东南立面 southeast elevation 西北立面 northwest elevation

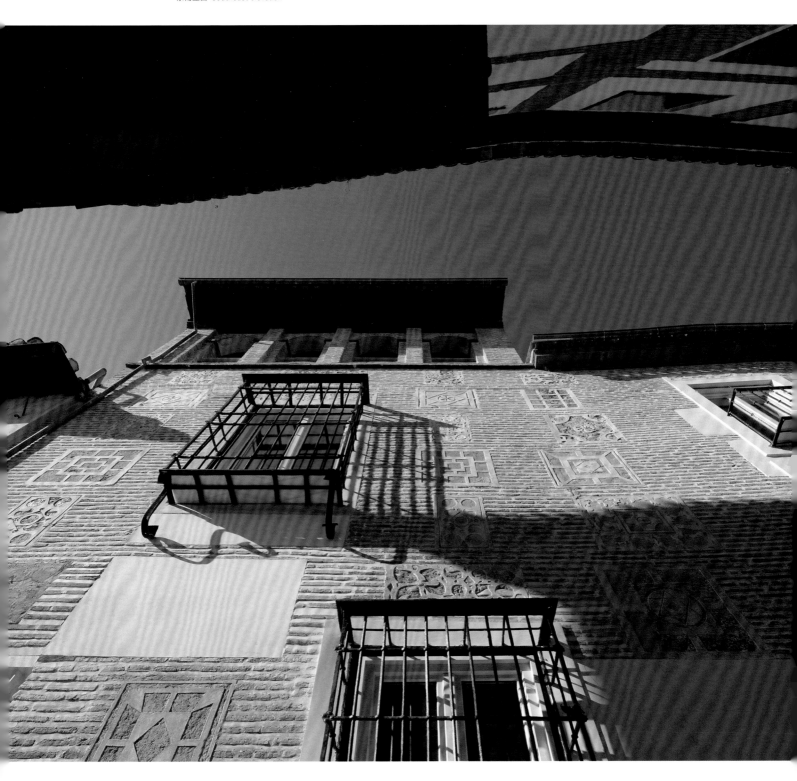

项目名称：Casa Zayas
地点：Cuesta de San Gregorio 13, 41002 Granada, Spain
建筑师：Ubaldo García Torrente, Marisol García Torrente
合作商：Isabel Mota Pernías, Antonio Lozano
施工方：Manuel Miranda Rojas. S.L.
赞助商：Oficina Rehabilitación Bajo Albaicín. EPSA. Junta de Andalucía, Fundación Zayas, Manuel Miranda Rojas. S.L.
甲方：Fundación Zayas
总楼面面积：707.06m²
设计时间：2010.4
竣工时间：2012.1
摄影师：©Fernando Alda

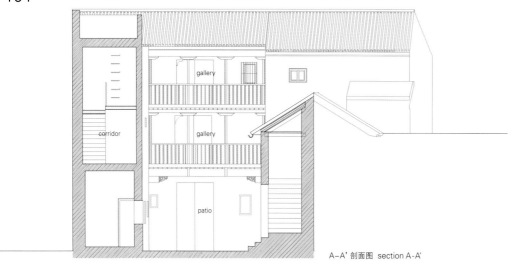

A-A' 剖面图 section A-A'

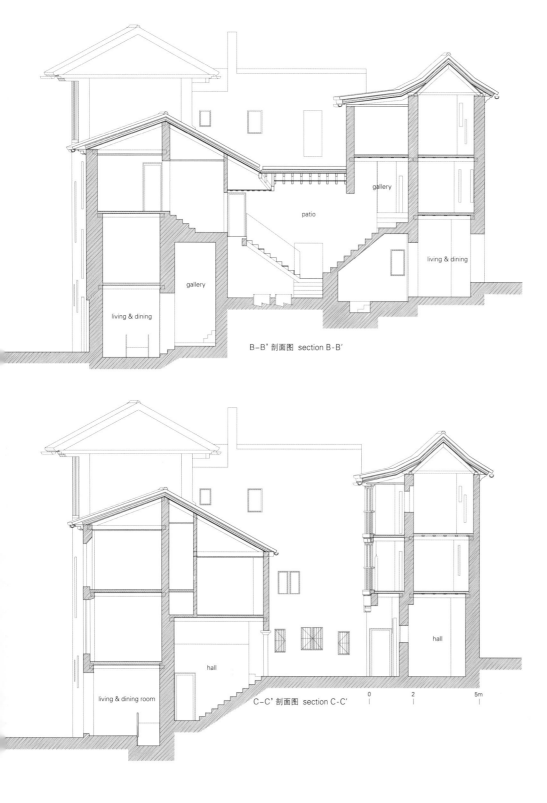

B-B' 剖面图 section B-B'

C-C' 剖面图 section C-C'

ousness is in its dual concept of interior-exterior and public-private... and versatileness in that it was created as a circulation space protected from rain, which is both a meeting place and neighborhood relationships, light sensor, air cool or heat, small garden, etc. A very sensitive area can capture perfectly the situation of its inhabitants. The gallery defines the courtyard, the gap.

The court focused on the plot is the social representation. The permeability between them allows adequate ventilation of the house. Ultimately the gallery and courtyard are the soul of the neighborhood, a vivid portrait of everyday life.

The Zayas house was one more example of noble residence Albaicín that with time and for various reasons, was renting piecewise other than denizens outside the household.

Today the building, after the intervention has become one, but inhabited by eight neighbors, all low-rent elderly, by decision of the former owner who gives to that end. Thanks to its flexibility, it is adaptabe to change according to use, property, neighborhood, dominant taste, etc.

An architectural potential, whose spatial structure, functional, light, formal, etc. is always open to redefinition and redeployment according to the criteria of value each time.

The building is reduced to a series of horizontal and vertical planes, walls and floors that shape the inhabitable areas: the courtyard is the void, the absence interior, the galleries as ambiguous limits.

Housing is abstract, organic and contradictory. Private enclosures for housing are geometrically pure and functional specialization, internal divisions disappear. Typically, a single area or a series of two or three homogeneously united by voids, usually without carpentry. All of these spaces are used at random and flexible, depending on the family unit or the specific circumstances of each clan that change throughout the long life. Specific applications are solved with furniture.

Both galleries and courtyards have public use, allowing relations between neighbors. The indoor-outdoor relationship becomes fluid, dynamic and rich: essential for a happy coexistence.

García Torrente Arquitectos

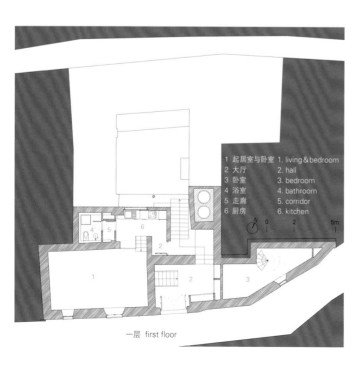

一层 first floor

1 起居室与卧室　1. living & bedroom
2 大厅　2. hall
3 卧室　3. bedroom
4 浴室　4. bathroom
5 走廊　5. corridor
6 厨房　6. kitchen

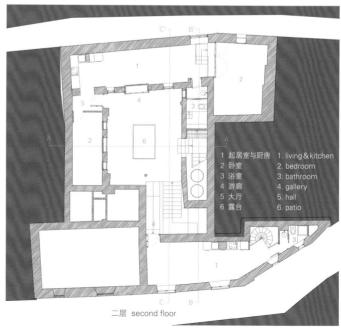

二层 second floor

1 起居室与厨房　1. living & kitchen
2 卧室　2. bedroom
3 浴室　3. bathroom
4 游廊　4. gallery
5 大厅　5. hall
6 露台　6. patio

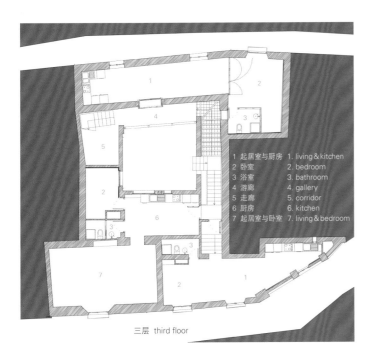

1 起居室与厨房	1. living & kitchen
2 卧室	2. bedroom
3 浴室	3. bathroom
4 游廊	4. gallery
5 走廊	5. corridor
6 厨房	6. kitchen
7 起居室与卧室	7. living & bedroom

三层 third floor

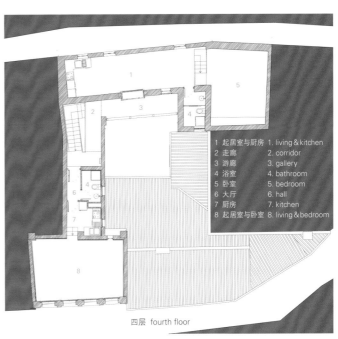

1 起居室与厨房	1. living & kitchen
2 走廊	2. corridor
3 游廊	3. gallery
4 浴室	4. bathroom
5 卧室	5. bedroom
6 大厅	6. hall
7 厨房	7. kitchen
8 起居室与卧室	8. living & bedroom

四层 fourth floor

潜望镜式住宅

C+ Arquitectos

潜望镜式住宅原是一处废墟，位于西班牙奥拓安普达的一个受保护的村落中。建筑师对其进行了修复，从而将其改建成度假住所。

复原的石墙上加盖宜居的屋顶，界定出了建筑的体量以及外围护结构。室内的设备安置在石墙上，作为双层表皮，以释放出中间的空间，这样一来，中间可以重新加以改造，以发挥不同的用途和功能，例如作为足球场、夜店或者冬季暖巢。

两个露台利用太阳光和风景打造出几何立体感，可以使黑暗的室内透光、通风，并使人们能够欣赏到室外的景观，宛如家用万花筒式的潜望镜。

从小处着眼，石墙上布满苔藓，镜面玻璃映射出的真实风景画面以及屋顶上可以欣赏到的绮丽风光，大自然的三种境界在这里得到完美结合。

从建筑学角度而言，这个房屋没有进出的道路，因此，所需材料都靠人工搬运和手动操作，包括调整尺寸和改动建筑体系。详细施工策略如下：

— 另类"郊区"感：项目提案中强调要尽可能保持现有遗迹，严格遵守建筑保护条例。与庸俗保守主义不同，施工过程中房屋原有的奇特之处都被保留下来，包括长苔藓的石墙、瓦片墙等，意在减少人为干预，并营造出一种新的材料特性。

— 搭建理想的建筑：讨论的核心不是地区或特定的材料，而是根据甲方的意愿采用具有象征意义的材料。房屋改建的过程犹如不同愿望具体付诸实践的过程：搭建森林里的小木屋、修建池塘、框式风景以及营造树木的味道等。

— 发挥持久功能：为了充分实现房屋的灵活性，空间被打造得像一个容器一样可以自如使用。设备被隐藏起来，家具和格挡可自由移动，房间没有规定具体用途，亦无专门的主人，房屋内空间分界不明显，从而提升了房屋的多功能用途。整个室内改造完给人一种游牧民族的气息。

— 技术与景观的运用：露台的几何形状可以调控自然光线和通风，同时还能通过反射来捕捉到室外的景观。

它们变成了万花筒，模糊了该住宅室内与室外的关系，营造出了一种使人仿佛置身于深山、树林与蓝天之中的环境。

— 抹去建筑痕迹：因为房屋外观受保护，辨识度不高，所以这所房屋没有立面或战略性视野。相反，室内是重点，通过各种反射效果和建筑活动打造而成。

— 甲方参与：房屋改建过程中遵从所有甲方的要求，考虑每个家庭成员、客人、爱好、愿望、承包商、地理位置、景观、天气和风等因素，竣工的房屋不仅满足上述要求，还利用各种社会技术手段实现相互间的自然过渡。

— 小巧并具有可持续性：在房屋改造过程中处处体现小巧，石墙、瓷砖、光滑的水泥和其他材料被无数次触摸和使用。每个角落都自成一体发挥作用。从另一个角度来看，每种策略中都体现着可持续理念，例如可再继承、太阳光和通风的被动式控制、没有挖掘环节、利用可再生能源或使用当地材料与当地供应商合作。

潜望镜式住宅融周边景致、甲方及其意愿为一体，各得其所。

Periscope House

The periscope house is a ruin rehabilitation of a protected village at the Alto Ampurdá, in Spain, to become a holiday residency.

An inhabitable roof lays on top of the recovered stone walls defining the volume and envelope of the house. In the interior the services are attached to the stone walls as a double skin to liberate the central space, which can be reconfigured in consonance with the different occupancies and uses desired: as a football field, night club or winter nest.

Two patios, whose geometry is shaped by sunlight and views, introduce light, ventilation and landscapes to the blind interior, becoming kaleidoscopic domestic periscopes.

Three scales of nature are then implemented: the microscopic of the mosses of the stone walls, the virtual of the landscape reflections in the mirror glass and the spectacular views of the surroundings that can be enjoyed at the roof.

From a constructive point of view the house does not have road access, and therefore all materials required hand transport and manipulation, including conditioning dimensions and constructive systems.

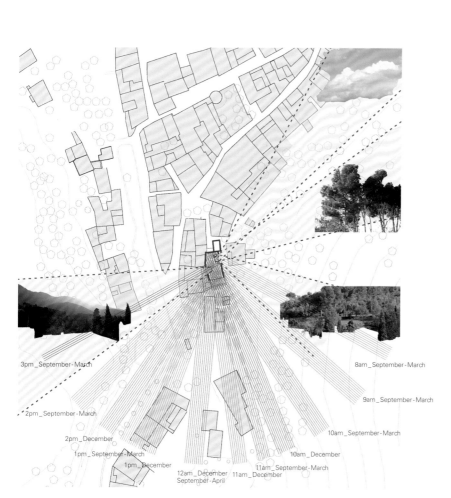

The strategies that articulate the project are:
- Another "rural": The proposal is to keep as much as possible the existing ruin and to take the conservation regulations as rules of the game. In contrast with a kitch conservationism all its singularities are incorporated: the mosses of the stones, the tilted walls, etc, to reduce the intervention and construct a new material identity.
- Construction of desires: The brief has not been articulated around areas or material specifications, but around symbolic material from the client´s fictions. The house is then built as a compendium of built desires with their different specificities: a hut in the woods, a pond, framed landscapes, the smell of wood…
- Continuous performativity: To stimulate the requested flexibility the space is built as a container to be appropriated. The services are hidden, the furniture and divisions are mobile and rooms don´t have name or owner, producing a spatial in-definition that promotes multiple ways of being. The domestic is understood as an action with nomad identity.
- Appropriation of technologies and landscapes: The geometry of the patios allows sunlight and ventilation control, and at the same time captures exterior landscapes through reflections. They become then kaleidoscopic devices that blur the relationship between the interior and the exterior of the house, produc-

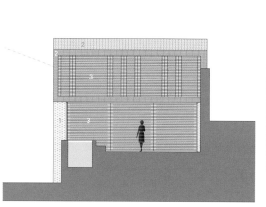

南立面 south elevation

北立面 north elevation

1. existing stone wall
2. flat and curved old ceramic finish
3. pine wood shades
4. steel handrail
5. rebuilt stone wall
6. mirror glass with aluminium frames

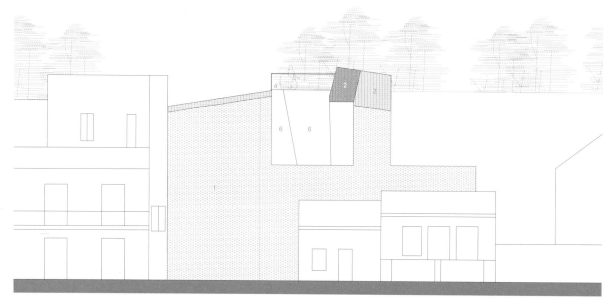

东立面 east elevation

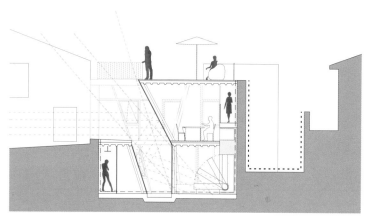

A-A' 剖面图 section A-A'

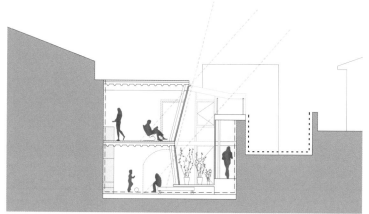

B-B' 剖面图 section B-B'

C-C' 剖面图 section C-C'

ing situations in which one is simultaneously in the mountains, the trees and the sky.
- Architecture as a disappearing strategy: As the exterior is protected and doesn't have good visibility, the project doesn't have facades or a strategic point of view. Instead, everything is interior, built through reflections and activity.
- Clients integration: The house is conformed as a clients' mediation space: each family member, their guests, their hobbies, their desires, the contractor, the location, the landscape, the weather, the wind.. all have their representation and are linked through various sociotechnic devices.
- Small and sustainable: "The small" is exploited: stone walls, ceramic tiles, sparkling concrete and other materials are touched and inhabited. Each corner is a micro space that can be activated. From another perspective, every strategy can be read in sustainable terms, as the heritage recuperation, the sunlight and ventilation passive control, the lack of excavation, the incorporation of renewable energies or the use of local materials and providers. The periscope house is therefore constituted as an invisible shaker of landscapes, clients and desires, which is activated in each use.

C+ Arquitectos

项目名称: Periscope House
地点: La Selva de Mar, Girona, Spain
建筑师: C+ Arquitectos
技术建筑师: Agustí Vidal
结构工程师: Francisco Poza
承包商: Constructora de l'Empordá Juliá Turrá
设备: Nieves Plaza
总楼面面积: 209m²
竣工时间: 2011
摄影师: ©Miguel de Guzmán

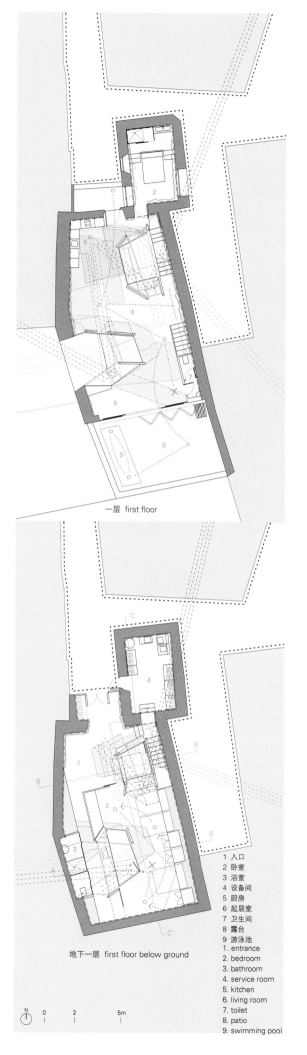

一层 first floor

地下一层 first floor below ground

1. 入口
2. 卧室
3. 浴室
4. 设备间
5. 厨房
6. 起居室
7. 卫生间
8. 露台
9. 游泳池

1. entrance
2. bedroom
3. bathroom
4. service room
5. kitchen
6. living room
7. toilet
8. patio
9. swimming pool

1. existing rock
2. pebbles e=20cm
3. concrete slab e=30cm
4. steel armature
5. underfloor heating e=12cm
6. continuous concrete finish with sparkling steel shavings
7. stiffening plate e=10mm
8. existing stone wall e=70cm
9. steel bars 4×ø10
10. elastic joint
11. steel plate (250×250×15mm)
12. round steel column ø10cm
13. protection resin, transparent
14. steel+concrete slab (6+6)
15. steel profile IPN-140
16. insulation e=8cm
17. waterproofing membrane
18. old ceramic tiles e=3cm
19. steel handrail
20. slope mortar
21. T steel profile (15×50mm)
22. double glass, SOLARUX, with reflective finish and air chamber
23. aluminium joinery
24. concrete infill
25. folded aluminium plate

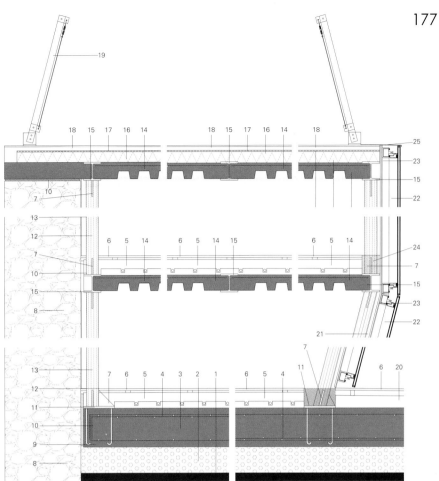

庭院详图 courtyard detail

Gonçalo Byrne+João Alexandre Góis+David Sinclair
Gonçalo Byrne is the founder and senior CEO of Gonçalo Byrne Arquitectos. For the last 35 years his works have been internationally recognized, prized and exhibited for accolades on projects in many countries. Has a teaching career primarily as visiting professor in several universities, including Coimbra, Lausanne, Venice, Mendrisio, Leuven, Harvard and Pamplona. João Alexandre Góis was born in Lisbon, Portugal in 1972. Received a master degree of architecture from the University Lusíada of Lisbon in 1999. Currently performs his professional activities in his own studio, Jag Architects in Lisbon.
David Sinclair trained in Plymouth and London. Qualified at the Royal Institute of British Architects in 1976. During his studies he traveled overland to India, Turkey, Iran, Afghanistan and Pakistan and moved to Portugal in late 1976. Has been working in Portuguese practices for four years and co-founded St George e Sinclair Architects with Alan St George.

>>158

García Torrente Arquitectos
Was founded by Ubaldo García Torrente and Marisol García Torrente in Spain. Both graduated from ETSA Sevilla, department of Architecture. Ubaldo had worked from 1991 to 2001 as program coordinator for international cooperation. Since 2000, Marisol has been an associate professor in architectural project area at School of Architecture in Granada.

>>128

CO-AP
Is a multidisciplinary architectural design office established in 2004. Will Fung[right] completed his bachelor degree of architecture with 1st class honors at the University of New South Wales(UNSW) in 1998. Traveled to the Netherlands in 1999 joining Wiel Arets Architects. Returning to Sydney in late 2000, he was a project architect at Engelen Moore until 2003. Has been invited as guest critic of the School of Architecture at UNSW. Began part time design tutoring at the School of Architecture at University of Sydney in 2009. Tina Engelen[left] has traveled internationally during formative years on a regular basis with her Dutch parents who imported architectural designed product for their family business, "Dedece". Studied Interior design at Sydney College of the Arts from 1984 to 1987 leaving to take position as features editor at *Interior Design and Architecture*, a new magazine published by Herbert Yoma.

>>22

Max Dudler
Was born in Switzerland in 1949. Studied architecture in Städelschule Frankfurt and at the Academy of Arts in Berlin. Has been running his own agency with offices in London, Zürich and Frankfurt since 1992. Was appointed as professor of architecture class at the Kunstakademie Düsseldorf in 2004. His architecture stands for a rationalistic language combining modernist basic tenets with normative aesthetic and contextual guiding principles of an urban architecture. Responds to the present cultural situation not by striving for universalization, overcoming, destruction and historic new beginning, but by accepting the challenge of postmodern pluralism.

>>38

FORM / Kouichi Kimura Architects
Kouichi Kimura was born in Kusatsu, Japan in 1960. Graduated from Kyoto Art College in 1982 and established FORM/Kouichi Kimura Architects in 1994.

>>168

C+Arquitectos
Was founded by Nerea Calvillo in 2004. She received a master's degree in advanced architecture design from Columbia University in 2001 as Fulbright Fellow, working at the moment on her PhD. Holds an assistant professorship at the European University of Madrid since 2006 and at the University of Alicante since 2010. Is currently a Poiesis Fellow at the New York University.

>>30
A-cero
Joaquin Torres^{left}, the director of A-cero was born in Barcelona. Was interested in art and architecture particularly from the early age. So that he decided to study architecture in the University of Corunna. Shortly after finishing the degree in 1996, he started up in the Galician capital, an architecture studio that was co-founded by other classmates and himself. Among them, Rafael llamazares^{right} continues to work under the name of A-cero.

>>70
Rafael Moneo
Was born in Tudela in the province of Navarra in 1937. Obtained his architectural degree in 1961 from the ETSAM. In 1985 he was appointed as chairman of the architecture department of the Harvard Graduate School of Design(GSD Harvard University) and held the position till 1990. Was named Jopsep Lluis Sert Professor of GSD Harvard University in 1991. Has been awarded numerous distinctions such as the Pritzker Prize for architecture in 1996 and the Royal Gold Medal of the Royal Institute of British Architects in 2003. Continues his professional activities as architect, lecturer, critic and theoretician.

>>116
Bak Gordon Arquitectos
Was founded by Ricardo Bak Gordon in 2000. He was born in Lisbon in 1967, and was licensed in 1990 at the faculty of architecture, Technical University of Lisbon. Was a representative of Portugal at the Venice Biennale 2010 with Alvaro Siza, Carrilho da Graca and Aires Mateus. Is currently a visiting professor in the master in architecture at the instituto Superior Tecnico, Lisbon, and visiting teacher at the Camilo Jose Cela University, Madrid.

>>140
Traverso-Vighy Architetti
Was established by Giovanni Traverso and Paola Vighy in 1996. Both of them graduated in architecture from the University Iuav of Venice(IUAV) in 1994 and completed their studies at the Bartlett, University College London. Are currently teaching at the University of Florida(UFL) and Interior Design Institute-ISAI. Directing the "Daylight thinking" international course about sustainable design at the UFL.

>>88
Nuno Ribeiro Lopes Arquitectos
Nuno Ribeiro Lopes was born in 1954 in Póvoa de Varzim, Portugal. Completed his degree in architecture at the Porto Higher Education School of Fine Arts in 1977. Has been an assistant visiting professor at Évora University, department of Architecture from 2005 to 2009 and a visiting professor at Coimbra University from 2009 to 2010. Currently is an author of articles and a speaker of national and international seminars.

>>48
Lussi+Halter Partner AG
Was founded by Thomas Lussi^{left} and Remo Halter^{right} in 2008 following General partnership Lussi+Halter which is established in 1999. Both studied architecture at the ETH Zürich. Thomas Lussi is an expert in the master diploma for architecture at the Lucerne University of Applied Science and Arts from 2010. Remo Halter was committee for professional guild of architecture SIA Switzerland from 2000 to 2006. Most of the projects are the achievement of competitions.

Silvio Carta
Is an architect and critic based in Rotterdam. Lives and works in the Netherlands, Spain and Italy, where he regularly writes reviews and critical essays about architecture and landscape for a diverse group of architecture magazines, newspapers and other media. Founded the Critical Agency™ | Europe in 2009.

Aldo Vanini
Practices in the fields of architecture and planning. Had many of his works published in various qualified international magazines. Is a member of regional and local government boards, involved in architectural and planning researches. One of his most important research interests is the conversion of abandoned mining sites in Sardini.

>>102

Alexandre Burmester Arquitectos Associados
Was founded by Alexandre Burmester in 1992. He has 30-year experience in architecture and urban design. Has ever worked as assistant designer of Architect Severiano Mário Porto in Rio de Janeiro in 1980. Since 1982 has worked as freelancer in his own Porto office.

>>150

Studio Inches Architettura
Was founded by Matteo Inches in 2010. He was born in 1984 and received a master's degree in architecture at the Academy of Architecture in Mendrisio in 2009. Has been teaching at the same school since 2011.

>>56

Stanton Williams Architects
Was founded by Alan Stanton and Paul Williams in 1985.
Paul Williams from the left has been a visiting critic and lectured at several universities and institutes in the UK and abroad. Is currently an examiner at Birmingham University School of Architecture. Patric Richard is passionate about the relationship between landscape and architecture, the urban realm and the role of architecture as a catalyst for regeneration. Alan Stanton has been a vice president of the architectural association council. Has lectured extensively in this country and abroad. Is currently an external examiner at Greenwich University. Believes that a great architecture is a profound understanding of space, and that context and heritage are creative forces in design. Peter Murray studied architecture at Melbourne University and the Polytechnic of Central London. Gavin Henderson has been a visiting critic and lecturer at a number of UK schools of architecture. Has a particular interest in the development and culture of cities and the role of public space.

C3, Issue 2012.10
All Rights Reserved. Authorized translation from the Korean-English language edition published by C3 Publishing Co., Seoul.
© 2012大连理工大学出版社
著作权合同登记06-2012年第234号

版权所有·侵权必究

图书在版编目(CIP)数据

建筑谱系传承：汉英对照 / 韩国C3出版公社编；于风军等译. —大连：大连理工大学出版社，2012.11
书名原文：C3:Genealogical Reasoning
ISBN 978-7-5611-7461-6

Ⅰ.①建… Ⅱ.①韩…②于… Ⅲ.①建筑设计－汉、英 Ⅳ.①TU2

中国版本图书馆CIP数据核字(2012)第285110号

出版发行：大连理工大学出版社
　　　　　（地址：大连市软件园路80号　邮编：116023）
印　　刷：精一印刷(深圳)有限公司
幅面尺寸：225mm×300mm
印　张：11.5
出版时间：2012年11月第1版
印刷时间：2012年11月第1次印刷
出 版 人：金英伟
统　筹：房　磊
责任编辑：张昕焱
封面设计：王志峰
责任校对：张媛媛

书　　号：ISBN 978-7-5611-7461-6
定　　价：228.00元

发　行：0411-84708842
传　真：0411-84701466
E-mail：12282980@qq.com
URL：http://www.dutp.cn